Structural Optimization with Uncertainties

SOLID MECHANICS AND ITS APPLICATIONS
Volume 162

Series Editors: G.M.L. GLADWELL
            Department of Civil Engineering
            University of Waterloo
            Waterloo, Ontario, Canada N2L 3GI

Aims and Scope of the Series

The fundamental questions arising in mechanics are: *Why?*, *How?*, and *How much?* The aim of this series is to provide lucid accounts written by authoritative researchers giving vision and insight in answering these questions on the subject of mechanics as it relates to solids.

The scope of the series covers the entire spectrum of solid mechanics. Thus it includes the foundation of mechanics; variational formulations; computational mechanics; statics, kinematics and dynamics of rigid and elastic bodies: vibrations of solids and structures; dynamical systems and chaos; the theories of elasticity, plasticity and viscoelasticity; composite materials; rods, beams, shells and membranes; structural control and stability; soils, rocks and geomechanics; fracture; tribology; experimental mechanics; biomechanics and machine design.

The median level of presentation is the first year graduate student. Some texts are monographs defining the current state of the field; others are accessible to final year undergraduates; but essentially the emphasis is on readability and clarity.

For other titles published in this series, go to
www.springer.com/series/6557

N.V. Banichuk · P.J. Neittaanmäki

Structural Optimization
with Uncertainties

 Springer

N.V. Banichuk
Inst. Problems in Mechanics
Russian Academy of Sciences
Prospect Vernadskogo 101
Moskva
Russia 117526
banichuk@ipmnet.ru

P.J. Neittaanmäki
Dept. Mathematical
Information Technology
University of Jyväskylä
FI-40351 Jyväskylä
Finland
pn@mit.jyu.fi

ISSN 0925-0042
ISBN 978-94-007-3111-0   e-ISBN 978-90-481-2518-0
DOI 10.1007/978-90-481-2518-0
Springer Dordrecht Heidelberg London New York

Springer is part of Springer Science+Business Media (www.springer.com)

Preface

Structural optimization is currently attracting considerable attention. Interest in research in optimal design has grown in connection with the rapid development of aeronautical and space technologies, shipbuilding, and design of precision machinery. A special field in these investigations is devoted to structural optimization with incomplete information (incomplete data). The importance of these investigations is explained as follows. The conventional theory of optimal structural design assumes precise knowledge of material parameters, including damage characteristics and loadings applied to the structure. In practice such precise knowledge is seldom available. Thus, it is important to be able to predict the sensitivity of a designed structure to random fluctuations in the environment and to variations in the material properties. To design reliable structures it is necessary to apply the so-called guaranteed approach, based on a "worst case scenario" or a more optimistic probabilistic approach, if we have additional statistical data.

Problems of optimal design with incomplete information also have considerable theoretical importance. The introduction and investigations into new types of mathematical problems are interesting in themselves. Note that some game-theoretical optimization problems arise for which there are no systematic techniques of investigation.

This monograph is devoted to the exposition of new ways of formulating and solving problems of structural optimization with incomplete information. We recall some research results concerning the optimum shape and structural properties of bodies subjected to external loadings. We study optimal design with incomplete information, accounting for the interaction between the structure and its environment, properties of materials, existence of initial damages and damage accumulation. This study is devoted to overcoming mathematical difficulties caused by local functionals. Most of the attention of the book is devoted to the minimax approach using worst case scenario, i.e. the guaranteed approach. But a probabilistic approach that does not guarantee the result is also described in the monograph, because it gives more "optimistic" results. Also, a mixed probabilistic-guaranteed approach is discussed and applied.

The authors offer a systematic and careful development of many aspects of structural optimization with incomplete information, particularly for rods, columns, beams, plates and shells. Some of the results are new and some have appeared only in specialized journals, or as proceedings of conferences. Some aspects of the theory presented here, such as shape optimization of beams, plates and shells with uncertainties, crack positioning and orientations, and optimization of structural elements with uncertain material properties, have not been considered before to any extent. Important new results relate to optimization of structures with a longevity constraint. Damage accumulation is modeled using Paris' law of fatigue cracks growth when the solid body is under cyclic loading. Incompleteness of information arises from uncertainties in initial crack length, orientation and positioning.

Our treatment is classical, i.e. it employs classical analysis: classical calculus of variations, functional analysis, the minimax approach of game theory and inequality analysis, and probabilistic approaches. We derive many results that are of interest to practical structural engineers, such as optimization with incomplete information concerning loads (which is the case in a great majority of practical design considerations and is of obvious interest in the design of aircraft, buildings, turbines, etc.).

This monograph contains material that has not been discussed elsewhere. However, there are other points of view concerning the formulation and solutions of stochastic problems. In this sense the monograph complements other developments in probabilistic optimal design. Especially it is concerned with the mixed probabilistic-guaranteed approach to structural optimization with uncertainties, and optimal design based on modern fracture mechanics. These new research directions were emphasized in a series of recent publications by the authors and their colleagues and are presented in this monograph.

The present monograph consists of three parts and 17 chapters. The first part is introductory; it contains two chapters (Chapters 1 and 2) and presents introduction, main notions and exposition of prototype problems; it also discusses the advantages and disadvantages of the various approaches.

Part II is devoted to optimal structural design with incomplete information, treated in the framework of the guaranteed approach. It contains 11 chapters. Chapter 3 describes various types of uncertainties, discusses Worst Case Scenario and presents a general transformation scheme that, in principle, allows one to reduce the original optimization problems with incomplete information to conventional problems of optimal structural design. A wide class of optimal design problems for thin-walled structures with uncertainties in loading conditions is considered in Chapters 3 and 4. In Chapter 5 we consider uncertainties of quasi-brittle fracture, and optimal design of structure with cracks. We describe some basic relations of fracture mechanics, model assumptions, and formulate optimization problems. Chapter 6 is devoted to optimal design of beams and plates taking into account fracture mechanics constraints and uncertainties in crack length and position. In Chapter 7 we present optimal design of axisymmetric shells with uncertainties in crack position and orientation, and find the optimal thickness distributions for toroidal, conical and spherical shells. Simultaneous optimization of the shape and thickness distribution of pressure vessels made of quasi-brittle materials is described in Chapter 8. Maximization of mass effectiveness of axisymmetric pressure vessels is discussed in Chapter 9. Optimal design of shells subjected to gravity forces and snow loading, and genetic algorithms for optimal shape and thickness distribution are presented in Chapter 10. Uncertainties in damage characteristics, and longevity constraints are discussed in Chapter 11. Modeling of cyclic loading taking into account the growth of initial cracks, worst case scenario and also analysis of minimax problems are presented in this chapter and in Chapter 12. Transformation of the optimization problems with incomplete information to conventional nonconvex problems and their solution by genetic algorithms are also included in Chapters 11 and 12. In Chapter 13 we present some results concerning an important class of optimal design problems with uncertainties in material characteristics. Special attention is devoted here to optimization of structures made from a discrete set of materials. These problems are nonconvex and constitute a special field in the theory of optimal structural design; their solution is found with the help of genetic algorithms.

Part III is devoted to optimal design with uncertainties in the framework of probabilistic approach and mixed probabilistic-guaranteed approaches. It consists of four chapters (Chapters 14–17). Chapter 14 is introductory and contains some basic notions of probability theory. Chapter 15 describes probabilistic approaches for optimal structural design with incomplete information. Here we present optimal shape designs for beams when crack length or position is taken as the random variable. Application of the probabilistic approach with direct constraint on stress intensity factor is also included in this chapter. Probabilistic problems of beams and frames under longevity constraint are discussed in Chapter 16. In Chapter 17 we describe a mixed probabilistic-guaranteed approach to optimal design of structures with cracks, and apply this approach to optimization of toroidal and other shells loaded by axisymmetric forces.

The book is intended for graduate and postgraduate students, specialists in mechanics and applied mathematics, engineers, university professors, etc. It is partially based on courses of lectures delivered by the authors at the Faculty of Information Technology of the University of Jyväskylä, in the Ishlinsky Institute for Problems in Mechanics of the Russian Academy of Sciences, and in the Department of Mathematics and Informatics of the University of Cagliari. We trust that knowledge acquired after 3 years of advanced mathematical training at a college or university would be sufficient for understanding most of the book.

Many results that are described in the book were derived by the authors in cooperation with F.J. Barthold, S.Yu Ivanova, E.V. Makeev, M.M. Mäkelä, F. Ragnedda, M. Serra and A.V. Sinitsin. We have included advice and remarks that were given in our study of specific problems by B.L. Karihaloo, A.S. Kravchuk, A.K. Lubimov, V.A. Palmov and E. Stein. To all of these, we express our sincere gratitude.

The book is written with support from RFBR (grant No. 08–08–00025-a), RAS Program No. 13 ("Accumulation of damages, fracture and . . . "), Program of Support of Leading Scientific Schools (grant No. 169.2008.1) and Tekes' (Finnish Funding Agency for Technology and Innovation) MASI (Modelling and Simulation) Programme.

Finally, we are very grateful to Professor G. Gladwell for his valuable comments. We wish also to acknowledge E.N. Bezrukova and Marja-Leena Rantalainen for their considerable efforts in the preparation of the manuscript.

N.V. Banichuk
P.J. Neittaanmäki

Contents

Part I
Prototype Problems

The majority of problems in the classical theory of optimal structural design are considered within the framework of an approach in which it is assumed that the loads applied are known and that we have complete knowledge regarding structural materials and boundary conditions. To solve such problems we can apply conventional techniques of the calculus of variations and optimal control theory. Problems of optimal design with incomplete information (with uncertainties) are entirely different in their formulation and solution technique. We shall consider here the simplest typical examples with uncertainties and discuss various approaches to optimization problems with incomplete information.

Part I

Chapter 1
Guaranteed Approaches

1.1 Prototype Problem 1

Consider optimal design of a cantilever beam clamped at the origin of Cartesian coordinate system ($oxyz$) and loaded by static transverse forces q acting in the plane xy. We suppose that the beam in its natural unloaded state is placed along the x-axis and that it has a rectangular cross-section with height $h = h(x)$ and width $b = $ const. The beam has length l. The function $h = h(x)$, determining the shape of the beam, is the unknown variable and is to be found in the design optimization process.

Some details of the applied admissible loads are not known beforehand. We assume that the admissible loads act in the positive direction of the y-axis, and that the load resultant does not exceed a given value P_0, i.e.

$$q(x) \geq 0, \quad \int_0^l q(x)dx \leq P_0. \tag{1.1}$$

The inequalities (1.1) describe the set of the all admissible loads that can be applied to the beam. Some particular realizations of the load distribution $\{q_1(x), q_2(x), \ldots, q_i(x), \ldots\}$ are shown in Fig. 1.1.

It is required that for any load satisfying the conditions (1.1), the normal stress σ_x should satisfy the imposed strength criterion

$$|\sigma_x| \leq \sigma_0, \tag{1.2}$$

where σ_0 is a specified material constant and the stresses $\sigma_x = \sigma_x(x, \zeta)$ are calculated according the formulas

$$\sigma_x = \frac{M\zeta}{I}, \quad I = \frac{bh^3}{12}, \tag{1.3}$$

where $M = M(x)$ denotes the bending moment. The coordinate ζ varies in the interval

N.V. Banichuk and P.J. Neittaanmäki, *Structural Optimization with Uncertainties*, Solid Mechanics and Its Applications 162, DOI 10.1007/978-90-481-2518-0_1, © Springer Science+Business Media B.V. 2010

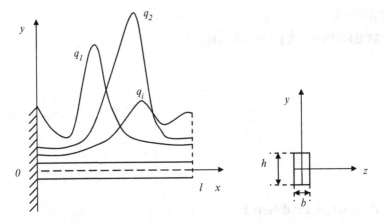

Fig. 1.1 Cantilever beam and admissible loads $q_1, q_2, \ldots, q_i, \ldots$

$$-\frac{h}{2} \leq \zeta \leq \frac{h}{2}. \tag{1.4}$$

The problem of optimizing the beam shape consists of finding a function $h = h(x)$ that for any admissible load $q = q(x)$, satisfying (1.1), satisfies the criterion (1.2) and minimizes the weight of the beam

$$J = \gamma b \int\limits_0^l h(x)dx \to \min_h, \tag{1.5}$$

under the additional geometric inequality

$$h(x) \geq h_o \geq 0, \tag{1.6}$$

where γ denotes the specific weight of the material and h_0 is a given problem parameter.

In order to express the criterion (1.2) as a constraint imposed on the desired function $h = h(x)$, we present necessary relations of the beam theory. The equation of equilibrium is written as

$$\frac{d^2 M}{dx^2} = -q. \tag{1.7}$$

The total bending moment $M(0)$ of the applied loads q with respect to the point $x = 0$ and the shear force dM/dx at this point are given by

$$M(0) = -\int\limits_0^l tq(t)dt, \tag{1.8}$$

$$\frac{dM}{dx}(0) = \int_0^l q(t)dt. \tag{1.9}$$

Using the expressions (1.8) and (1.9) as boundary conditions for Eq. (1.7) and performing integration, we will have the following expression for the distribution of bending moment:

$$M(x) = \int_x^l q(t)(x - t)dt. \tag{1.10}$$

Consider the properties of the bending moment distribution $M = M(x)$ as a function of applied loads $q = q(\xi)$, $0 \le \xi \le 1$, satisfying the conditions (1.1). From the formulas (1.1), (1.10) it follows that

$$M(x) \le 0, \ \ 0 \le x \le l. \tag{1.11}$$

Let us evaluate

$$M^{\max}(x) = \max_q |M(x)| \tag{1.12}$$

that represents, for fixed $x \in [0, l]$, the maximum of the modulus of the bending moment with respect to admissible realization of loads $q = q(x)$, $0 \le x \le l$, satisfying the conditions (1.1). We will prove the following property:

$$M^{\max}(x) = P_0(l - x), \ \ 0 \le x \le l, \tag{1.13}$$

and maximum of $|M(x)|$ for any x from the interval $[0, 1]$ is realized for a point load P_0 acting at the point $x = 1$, i.e. $q(x) = P_0\delta(x - l)$, where δ denotes the Dirac delta function. It follows from the estimates

$$|M(x)| \le \max_t |x - t| \int_x^l q(t)dt \le (l - x)P_0, \tag{1.14}$$

where the maximum with respect to t is evaluated for $x \le t \le l$. We also observe that for the admissible load $q(t) = P_0\delta(t - l)$, the value of the bending moment is

$$M(x) = P_0(l - x). \tag{1.15}$$

Combining this equality with the estimates given above, we see that the equality (1.13) is true. Various admissible realizations of bending moments corresponding to point loads P_0 applied at the points $x = x_1, x = x_2, \ldots, x = l$ ($q = P_0\delta(x - x_i)$) are shown in Fig. 1.2 by straight solid and dashed lines.

Fig. 1.2 Admissible
realizations of the bending
moments

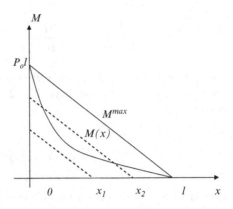

Using presented estimates, we evaluate the maximum normal stresses as

$$|\sigma_x(x)| \leq \max_{\zeta} \max_{q} \left| \frac{M\zeta}{I} \right| = \frac{6P_0(l-x)}{bh^2}. \tag{1.16}$$

Here the maximum with respect to ζ is determined under constraints (1.4) and is achieved for upper and lower fibers of the rectangular cross-section of the beam, i.e. for $\zeta = \pm h/2$.

Substitution of (1.16) into (1.2) permits us to express the original strength criterion in the form of a constraint imposed on the maximum stress σ_x^{max}:

$$\sigma_x^{max} = \frac{6P_0(l-x)}{bh^2} \leq \sigma_0. \tag{1.17}$$

Using the strength criterion (1.17) and the geometric condition (1.6), we obtain the following explicit constraint for the desired design variable:

$$h(x) \geq \max\{h_0, H(x)\}, \tag{1.18}$$

$$H(x) = \left[\frac{6M^{max}(x)}{b\sigma_0} \right]^{1/2} = \left[\frac{6P_0(1-x)}{b\sigma_0} \right]^{1/2} \tag{1.19}$$

For each $x \in [0, l]$ maximum operation in (1.18) means to find the maximum of the two quantities written in the braces.

It follows directly from (1.5), (1.18), (1.19) that the function

$$h_*(x) = \max\{h_0, H(x)\}, \quad 0 \leq x \leq l \tag{1.20}$$

satisfies the constraint (1.18), (1.19) (strength and geometric requirements) and assigns a minimum to the integral (1.5). Consequently it is a solution of our optimization problem. Optimal thickness distribution $h_*(x)$ is represented in Fig. 1.3.

Fig. 1.3 Optimal shape of
the beam

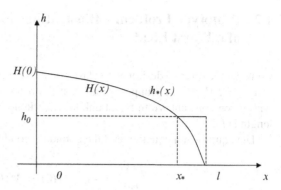

The optimal solution (1.20) can also be represented in the following form:

$$h_*(x) = H(x), \quad 0 \le x \le x_*, \tag{1.21}$$
$$h_*(x) = h_0, \quad x_* \le x \le l \tag{1.22}$$

in the case when

$$h_0 \le \sqrt{\frac{6Pl}{b\sigma_0}}.$$

For other case

$$h_0 \ge \sqrt{\frac{6Pl}{b\sigma_0}}$$

the optimal solution is given by (1.22) with $x_* = 0$. The magnitude of x_* can be found by using the assumption that $h_*(x)$ is a continuous function at $x = x_*$. We have

$$x_* = l - \frac{b\sigma_0 h_0^2}{6P_0}. \tag{1.23}$$

Thus, in this case there is a worst load in the set of admissible loads (1.1). This load is a point transverse force applied to the free end of the cantilever, i.e.

$$q_{\text{worst}} = P_0 \delta(l - x).$$

The optimization problem can be solved for the worst case scenario. In accordance with this scenario we consider only the worst load and determine the corresponding bending moment (1.13) and maximum stresses (1.17). The strength constraint is represented in the pure "deterministic" form (see the right-hand side in (1.17)) and the original optimization problem with incomplete information, concerning applied loads, is transformed to a conventional optimization problem (1.5), (1.18), (1.19), minimization of the integral (1.5) under the geometrical constraint (1.18), (1.19).

1.2 Prototype Problem 2 Illustrating Nonexistence of a Worst Load

Consider the optimal design of an elastic beam simply supported at the points $x = 0$ and $x = l$ (Fig. 1.4). Suppose again that the beam has a rectangular cross-section with given constant width b and unknown variable thickness $h = h(x)$. The beam length is l.

The equilibrium equation and the boundary conditions are given by

$$\frac{d^2M}{dx^2} = -q, \quad M(0) = M(l) = 0. \tag{1.24}$$

The set of admissible loads $q = q(x)$ applied to the beam is described by the inequalities (1.1), and the strength criterion has the form (1.2), (1.3). Taking the thickness distribution $h = h(x)$ $(0 \leq x \leq l)$ as an unknown design variable, we consider the problem of minimization of the integral (1.5) under strength and geometric constraints (1.2), (1.3), (1.6).

Before we attempt to solve this problem we need to explain some properties of the function $M = M(x)$ that we shall need in our study. Integrating the equilibrium equation with the indicated boundary conditions (1.24), we obtain the solution of the boundary-value problem (1.24) in the form

$$M(x) = \int_0^l K(x,t)q(t)dt, \ 0 \leq x \leq l, \tag{1.25}$$

$$K(x,t) = t\left(1 - \frac{x}{l}\right), \ 0 \leq t \leq x,$$

$$K(x,t) = x\left(1 - \frac{t}{l}\right), \ x \leq t \leq l.$$

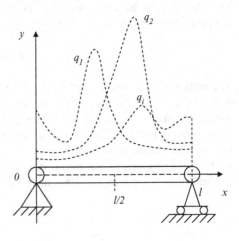

Fig. 1.4 Simply supported beam and admissible loads $q_1, q_2, \ldots, q_i, \ldots$

Since the functions $K(x, t)$ and $q(x)$ are positive, it follows that $M(x) \geq 0$ for $0 \leq x \leq l$. We fix a point $x \in [0, l]$ and consider the set of values that can be attained by $M = M(x)$ for all possible admissible loads $q = q(t)$, $0 \leq t \leq l$, satisfying the inequalities (1.1). We denote by

$$M^{\max}(x) = \max_q M(x) \tag{1.26}$$

the maximum value of the bending moment and prove that

$$M^{\max}(x) = P_0 x \left(1 - \frac{x}{l}\right), \quad 0 \leq x \leq l. \tag{1.27}$$

To prove this assertion, we use the expression (1.25) and make the following estimates:

$$M(x) \leq \max_q \int_0^l K(x, t) q(t) dt \leq \int_0^l \max_q \left[K(x, t) q(t)\right] dt. \tag{1.28}$$

Taking into account the properties of δ-function and that the maximum $K(x, t)$ is attained for $t = x$, we will have

$$\int_0^l \max_q \left[K(x, t) q(t)\right] dt \leq \max_t K(x, t) \int_0^l q(\xi) d\xi = P_0 x \left(1 - \frac{x}{l}\right) \equiv M^{\max}(x).$$

$$\tag{1.29}$$

The maximum with respect to t is evaluated for $0 \leq t \leq l$. We also observe that for the admissible load

$$q(t) = P_0 \delta(t - x) \tag{1.30}$$

the value of the bending moment, as it is seen from (1.25), is

$$M(x) = (M(x))_{q = P_0 \delta(t - x)} = P_0 x \left(1 - \frac{x}{l}\right). \tag{1.31}$$

Combining this with the estimates given above, we see that the equality (1.27) is true.

It is important to note that for each fixed x the load q that realizes the maximum of $M(x)$ is the point force (δ-function) and this force acts at the same point. Thus for various points x we need to take different realization of loads from (1.1) to achieve the maximum value of the bending moment. Distribution of bending moments $M_i = M_i(x)$ for each separate realization of $q_i = P_0 \delta(x_i - t)$ is shown in Fig. 1.5 by dashed broken lines.

It is seen from Fig. 1.5 that the maximum value of each $M_i = M_i(x)$ is attained for $x = x_i$ when $q = q_i = P_0 \delta(x_i - t)$, i.e.

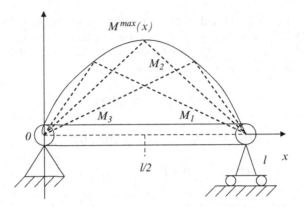

Fig. 1.5 Bending moments corresponding to the pointed admissible loads of the limit value P_0

$$(M_i(x))_{q=q_i} = M^{\max}(x_i). \tag{1.32}$$

Thus the enveloping curve $M^{\max}(x)$ (envelope), shown in Fig. 1.5 by continuous curve, consists of the points of maximum of the bending moment functions. Consequently any distribution of the bending moment $M = M(x)$, corresponding to a point force or other load distribution from (1.1), lies below the envelope, i.e.

$$M(x) \le M^{\max}(x). \tag{1.33}$$

For each admissible load distribution the rigorous equality in (1.33) can be achieved at not more than one point. Therefore, for each separate point x_0 $(0 < x_0 < l)$ the worst load is $P_0 \delta(x - x_0)$ and there is no worst load for the whole interval.

Normal stresses are estimated with the help of (1.2)–(1.4), (1.27). We have

$$\sigma_x(x) \le \max_{\zeta} \max_{q} \left| \frac{M\zeta}{I} \right| = \frac{6P_0 x}{bh^2} \left(1 - \frac{x}{l} \right) \equiv \sigma_x^{\max}. \tag{1.34}$$

Using a minimax approach based on the strength criterion (1.2) and estimation (1.34), we transform the original optimization problem with incomplete information, concerning applied load from the set (1.1), to the classical deterministic optimization problem of minimizing the integral (1.5) under the geometric constraint (1.6) and the inequality

$$h(x) \ge H(x) \equiv \left(\frac{6P_0 x}{b\sigma_0} \left(1 - \frac{x}{l} \right) \right)^{1/2} \tag{1.35}$$

arising from (1.2), (1.34). Consequently, the optimal thickness distribution has the form

$$h_*(x) = \max\{h_0, H(x)\} \tag{1.36}$$

shown in Fig. 1.6.

Fig. 1.6 Optimal thickness
distribution of simply
supported beam

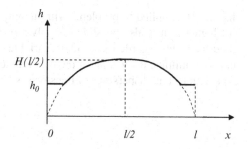

Note that the max operation in (1.36) means finding the greater of the two quantities in brackets. The optimal thickness distribution can be also written as

$$
h_*(x) = \begin{cases} h_0, & 0 \leq x \leq x_* \\ H(x), & x_* \leq x \leq l - x_* \\ h_0, & l - x_* \leq x \leq l \end{cases} \tag{1.37}
$$

if

$$
h_0 < \max_x H(x) = \left[\frac{3 P_0 l}{2 b \sigma_0} \right]^{1/2}. \tag{1.38}
$$

The parameter x_* in (1.37) is

$$
x_* = \frac{l}{2} - \sqrt{1 - \frac{2 b \sigma_0}{3 P_0 l} h_0^2}. \tag{1.39}
$$

If the inequality (1.38) is violated then the optimal thickness distribution is constant: $h_*(x) = h_0$.

In this section we applied guaranteed approach to formulation and solution of the beam optimization problems with incomplete information regarding loading conditions. In application of this approach we have to specify the set containing all admissible loads. The beam of a given shape is optimum if for any other beam having a smaller weight it is possible to find a load in the class of admissible loads for which some strength constraint is violated. As we observe in Sections 1.1 and 1.2, one of two possibilities emerges. There may exist a "worst load" (Section 1.1) for which the beam has a minimum weight, this weight being computed specifically for that load, while the constraint on strength is satisfied for this and all other admissible loads. Otherwise, the beam of a given shape is optimum for the class of admissible loads (Section 1.2), i.e. it is the solution of the original optimization problem, but a "worst load" does not exist and the optimum design for the class of admissible loads is not optimum for any specific load in the admissible class.

Note that the guaranteed approach may also be applied to problems with incomplete information concerning the boundary conditions or properties of structural material. We also remark that the minimax approach is not the only available one that may be applied to problems with incomplete information. As it will be shown in Chapter 2, it is also possible to apply a probabilistic approach, in which the applied loads are regarded as random variables with a given probability distribution and we minimize the weight of the beam under the constraints on the mathematical expectations and dispersions of the stresses.

Chapter 2
Probabilistic Uncertainties

In this book we will consider different type of uncertainties. In particular, we discuss optimal structural design problems with incomplete information taking into account that some problem parameters take random values with a given probability density. For example, for optimal design problems, considered in Part III in the framework of modern fracture mechanics, the role of random parameters is given by the sizes, positioning and orientations of cracks. In this chapter we will consider the simplest optimization problems with random parameters and discuss some possible approaches to these problems.

2.1 Prototype Problem 3: Moment Constraints

Consider some possible formulations of optimization problems with random parameters. To be specific, consider the optimal design of a simply supported beam having length l, rectangular cross-section with constant width b ($b = $ const) and variable thickness distribution $h = h(x)$ considered as the desired design variable. Suppose that the load

$$q = q(x, \xi), \ 0 \le x \le l \qquad (2.1)$$

is linear in one amplitude parameter ξ, i.e.

$$q(x, \xi) = \xi q_0(x), \qquad (2.2)$$

where $q_0(x) \ge 0$ is a given function. Load distributions are shown in Fig. 2.1 for different meanings of the parameter ξ.

The amplitude parameter ξ is considered here as a random variable with given probability density $f(\xi)$ and probability distribution function $F(\xi)$

$$\frac{dF(x)}{dx} = f(x) \qquad (2.3)$$

as shown in Fig. 2.2.

N.V. Banichuk and P.J. Neittaanmäki, *Structural Optimization with Uncertainties*, Solid
Mechanics and Its Applications 162, DOI 10.1007/978-90-481-2518-0_1,
© Springer Science+Business Media B.V. 2010

Fig. 2.1 Load distributions
with various realization of the
parameter ξ : $q_1 = \xi_1 q_0(x)$,
$q_2 = \xi_2 q_0(x), \ldots$

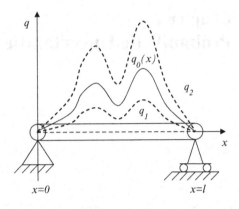

Fig. 2.2 Probability density
and distribution function

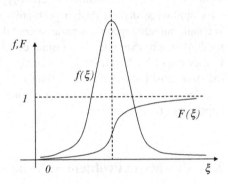

Taking into account that the normal stresses, arising from the bending process, achieve the maximum value at the external fibers of the rectangular cross-section of the beam, we will write the strength constraint as

$$\sigma_x^0\,(x, h(x), \xi) = \left| \frac{M(x, \xi)\zeta}{I(h(x))} \right|_{\substack{\zeta = \pm \frac{h(x)}{2} \\ x = 0}} = \frac{6M(x, \xi)}{bh^2(x)} \le \sigma_0, \qquad (2.4)$$

where $\sigma_0 > 0$ is a given positive constant.

The stress σ_x^0 depends implicitly (through the dependence of the bending moment M on the loads q) on the random variable ξ. Let us express $M(x, \xi)$ explicitly. To do this, we use (1.25), (2.2):

$$M(x, \xi) = \xi M_0(x),\ 0 \le x \le l, \qquad (2.5)$$

$$M_0(x) = \int_0^l K(x, t) q_0(t) dt, \qquad (2.6)$$

where

$$K = t\left(1 - \frac{x}{l}\right), \ 0 \leq t \leq x; \ K = x\left(1 - \frac{t}{l}\right), \ x \leq t \leq l.$$

The probability density $f(\xi)$ for random variable ξ permits us, at least in principle, to determine the moments of the random function σ_x^0, and in particular, its mathematical expectation (mean value) and dispersion (variance).

For each $x \in [0, l]$, the mean value of the normal stresses is estimated as

$$\widehat{\sigma}_x^0 = \int_0^\infty \sigma_x^0(x, \xi) f(\xi) d\xi \tag{2.7}$$

and the dispersion of $\sigma_x^0(x, \xi)$ is evaluated using the following expression:

$$D\left(\sigma_x^0\right) = \int_0^\infty \left(\sigma_x^0(x, \xi) - \widehat{\sigma}_x^0\right)^2 f(\xi) d\xi. \tag{2.8}$$

In what follows we will also use the following notation for functions of the independent variable x:

$$\widehat{\sigma}_x^0 = E(\sigma_x^0) = q_\varepsilon(x), \tag{2.9}$$

$$D(\sigma_x^0) = E\left(\left(\sigma_x^0 - \widehat{\sigma}_x^0\right)^2\right) = g_D(x), \tag{2.10}$$

where $E(\sigma_x^0)$ and $D(\sigma_x^0)$ are respectively the mathematical expectation and dispersion for the random variable.

These characteristics allow us to control the probability of violating the assigned constraint (2.4) on strength, and to study the following problem of optimal design, finding a thickness distribution $h = h(x)$ that satisfies the following system of inequalities:

$$E(\sigma_x^0) = g_E(x) \leq \sigma_0, \tag{2.11}$$

$$D(\sigma_x^0) = g_D(x) \leq \delta \tag{2.12}$$

and minimizes the functional $J = J(h)$ (the weight of the beam, represented by the expression (1.5)) under the additional geometric constraint (1.6). Here δ denotes a sufficiently small number which is supposed to be given. The inequalities (2.11) and (2.12) imply that some conditions must accompany a choice of the design variable h. The strength constraint given in (2.4) must be satisfied in "an average" sense, and the deviation of the quantity $\sigma_x^0 - \widehat{\sigma}_x^0$ must not exceed δ.

To find explicit constraints for the unknown thickness distribution imposed by the inequalities (2.11), (2.12) it is necessary to express the expectation and dispersion of σ_x^0 in terms of h. For this purpose we will use the expressions (2.4), (2.5) and

(2.7)–(2.10). Performing elementary transformations, we will have the following formula for the expectation:

$$E(\sigma_x^0) = \varphi(x)E(\xi) \equiv g_E(x),\qquad(2.13)$$

where

$$\varphi(x) = \frac{6M_0(x)}{bh^2(x)}, \quad E(\xi) = \int_0^\infty \xi f(\xi)d\xi.\qquad(2.14)$$

The dispersion is

$$D(\sigma_x^0) = \varphi^2(x)D(\xi) \equiv g_D(x),\qquad(2.15)$$

where

$$D(\xi) = \int_0^\infty (\xi - E(\xi))^2 f(\xi)d\xi = E(\xi^2) - E^2(\xi).\qquad(2.16)$$

Using the explicit expressions (2.13)–(2.16), we can transform the constraints (2.11), (2.12) to explicit inequalities imposed on the thickness distribution $h(x)$:

$$h(x) \geq \psi(x)\left(\frac{E(\xi)}{\sigma_0}\right)^{1/2}\qquad(2.17)$$

and

$$h(x) \geq \psi(x)\left(\frac{D(\xi)}{\delta}\right)^{1/4},\qquad(2.18)$$

where

$$\psi(x) = \sqrt{\frac{6M_0(x)}{b}}.\qquad(2.19)$$

Thus the original stochastic optimization problem has been transformed to a classical optimization problem of minimization of the integral (1.5) under the geometric constraints (1.6), (2.17)–(2.19). The thickness distribution

$$h_*(x) = \max\left\{h_0, \psi(x)\left(\frac{E(\xi)}{\sigma_0}\right)^{1/2}, \psi(x)\left(\frac{D(\xi)}{\delta}\right)^{1/4}\right\}\qquad(2.20)$$

represents the desired optimal solution. The expression for the optimal solution (2.20) can be transformed if we compare the second and the third terms in the right hand side of (2.20) (in braces): the optimal solution $h_*(x)$ is

$$h_*(x) = \max\left\{h_0, \psi(x)\left(\frac{E(\xi)}{\sigma_0}\right)^{1/2}\right\}\qquad(2.21)$$

if the problem parameters satisfy the inequality

$$\frac{E(\xi)}{\sqrt{D(\xi)}} \geq \frac{\sigma_0}{\sqrt{\delta}}.$$

(2.22)

Otherwise, when the inequality (2.23) is violated, the optimal solution is

$$h_*(x) = \max\left\{h_0, \psi(x)\left(\frac{D(\xi)}{\delta}\right)^{1/4}\right\}.$$

(2.23)

If

$$h_0 \geq \left(\frac{E(\xi)}{\sigma_0}\right)^{1/2} \max_x \ \psi(x)$$

(2.24)

and

$$h_0 \geq \left(\frac{D(\xi)}{\delta}\right)^{1/4} \max_x \ \psi(x),$$

(2.25)

then the optimal thickness distribution will be constant

$$h_*(x) = h_0, 0 \leq x \leq l.$$

(2.26)

2.2 Prototype Problem 4: Optimization with Direct Probabilistic Constraints

We now describe another probabilistic approach for the formulation and solution of structural optimization problems with incomplete information [BRS99]. Reconsider the minimizing the weight of a simply supported beam loaded by a random transverse force

$$q = q(x, \xi), \ 0 \leq x \leq l$$

(2.27)

with random parameter ξ and given probability density $f = f(\xi)$. The distribution of maximum normal stresses σ_x^0 is given by

$$\sigma_x^0(x, h(x), \xi) = \frac{6M(x, \xi)}{bh^2(x)} = \frac{6}{bh^2(x)} \int_0^l K(x, t) q(t, \xi) dt,$$

(2.28)

where $K(x, t)$ is as in (2.6). Suppose that $\sigma_x^0(x, h(x), \xi)$ is a monotonic function of ξ and the equation

$$\sigma_x^0(x, h(x), \xi) = \sigma_0$$

(2.29)

has a unique solution

$$\xi_0 = \xi_0(x, h(x), \sigma_0)$$

(2.30)

for fixed x and σ_0. Then for any x from the interval $[0, l]$ the probability P of satisfying the inequality

$$\sigma_x^0(x, h(x), \xi) \leq \sigma_0$$

can be estimated as

$$P\left\{\sigma_x^0(x, h(x), \xi) \leq \sigma_0\right\} = P\left\{\xi \leq \xi_0\right\}, \qquad (2.31)$$

where

$$P\left\{\xi \leq \xi_0\right\} = \int_0^{\xi_0(x, h(x), \sigma_0)} f(\xi)d\xi. \qquad (2.32)$$

Consequently, the probabilistic expression for the strength constraint (reliability constraint) can be written as

$$\min_x P\left\{\sigma_x^0(x, h(x), \xi) \leq \sigma_0\right\} = P\left\{\max_x \sigma_x^0(x, h(x), \xi) \leq \sigma_0\right\} \geq 1 - v \quad (2.33)$$

or in the following form:

$$\min_x \int_0^{\xi_0(x, h(x), \sigma_0)} f(\xi)d\xi \geq 1 - v. \qquad (2.34)$$

Consider the equation

$$F(\xi_m) = \int_0^{\xi_m} f(\xi)d\xi = 1 - v. \qquad (2.35)$$

Taking into account the positiveness of the probability density $f(\xi)$ and that the probability distribution (integral distribution)

$$F(\xi) = \int_0^{\xi} f(\xi)d\xi \qquad (2.36)$$

is a monotonically increasing function, we will use the unique solution of Eq. (2.35)

$$\xi_m = F^{-1}(\kappa), \quad \kappa = 1 - v \qquad (2.37)$$

and express the reliability condition (2.34) in the following form:

$$\xi_0(x, h(x), \sigma_0) \geq \xi_m. \qquad (2.38)$$

Suppose for concreteness, that the random load acting on the beam is given by (2.2) and, consequently, $\xi_0(x, h(x), \sigma_0)$, as is seen from (2.2), (2.4)–(2.6), (2.29), can be written as

$$\xi_0(x, h(x), \sigma_0) = \frac{b\sigma_0 h^2(x)}{6M_0(x)}. \tag{2.39}$$

Using the relations (2.38), (2.39) we transform the original probabilistic problem to a classical problem of minimization of the integral (1.5) under constraint (1.6), and the inequality

$$h^2(x) \geq \xi_m \left(\frac{6M_0(x)}{b\sigma_0} \right), \quad 0 \leq x \leq l. \tag{2.40}$$

The solution of the classical minimization problem can be represented in the following way:

$$h_*(x) = \max \left\{ h_0, \left[F^{-1}(1-v) \frac{6M_0(x)}{b\sigma_0} \right]^{1/2} \right\}. \tag{2.41}$$

If the problem parameters satisfy the condition

$$h_0^2 \geq \frac{6F^{-1}(1-v)}{b\sigma_0} \max_x M_0(x) \tag{2.42}$$

then the optimal thickness distribution $h_*(x)$ will be constant, i.e. $h_*(x) = h_0$.

Part II
Optimization in Frame of Guaranteed Approach

Chapter 3
Uncertainties and Worst Case Scenarios

3.1 Transformation Schemes

In this section we describe details of problem formulation in abstract form. Let the behavioral system of differential equations with boundary conditions, described the equilibrium of an elastic body, be of the operator form

$$L(u, h, q, \xi, \omega) = 0 \qquad (3.1)$$

and is written in the domain Ω in n-dimensional space with boundary $\Gamma = \partial\Omega$, where u, h, q are respectively the state variable, the design variable, the applied force and the functions ξ, ω characterize the material distribution along the body and the distribution of damages.

The type of load applied to the body is not known a priori. Besides the body is often made of materials which properties are known with some error. The parameters of damages, characterizing their sizes, location, shape and orientation (in case of crack appearance) are also can not be specified rigorously. Taking this into account we specify only the sets Λ_q, Λ_ξ and Λ_ω which contain all admissible loads, material properties and damage characteristics. We denote this by writing

$$q \in \Lambda_q, \qquad (3.2)$$

$$\xi \in \Lambda_\xi, \qquad (3.3)$$

$$\omega \in \Lambda_\omega. \qquad (3.4)$$

In formulating the problems of optimizing the shape of the body or its internal structure and topology we only consider admissible forces, material properties and damage characteristics defined by the constraints (3.2)–(3.4).

For given q, ξ, ω and h, we assume that the boundary-value problem (3.1), consisting of finding u, has a unique solution. The optimization problem consists of finding a design variable

$$h(x) \in H, \; x \in \Omega \qquad (3.5)$$

N.V. Banichuk and P.J. Neittaanmäki, *Structural Optimization with Uncertainties*, Solid
Mechanics and Its Applications 162, DOI 10.1007/978-90-481-2518-0_1,
© Springer Science+Business Media B.V. 2010

that minimizes the functional $J(h)$ (the weight or volume of the body or the cost of the structure) and satisfies, for arbitrary q, ξ and ω in (3.2)–(3.4), the strength constraints, written in abstract form as

$$\psi(x, u, h, q, \xi, \omega) \leq 0, \ x \in \Omega, \tag{3.6}$$

where ψ is a known vector-valued function and H, Ω are given domains. The constraints (3.6) are represented by a system of scalar inequalities. Note that because of the indeterminacy in the system, the problem belongs to game theory (we play games with nature). The correct formulation is attained if we will use a guaranteed or minimax approach based on the worst-case scenario (WCS).

We explain the main features of this approach. Denote a solution of the boundary value problem (3.1) by

$$u = u(x, h, q, \xi, \omega). \tag{3.7}$$

The dependence of u on h, q, ξ and ω is, in general, a functional dependence. Because of this, the inequality (3.6) can be considered as a system of functional inequalities

$$\Psi(x, h, q, \xi, \omega) \leq 0, \tag{3.8}$$

where

$$\Psi(x, h, q, \xi, \omega) \equiv \psi(x, u(x, h, q, \xi, \omega), h, q, \xi, \omega). \tag{3.9}$$

Denoting the components of the vector Ψ by Ψ_j, we consider the worst case when $\Psi_j(x, h, q, \xi, \omega)$ arrives at maxima with respect to $q \in \Lambda_q$, $\xi \in \Lambda_\xi$ and $\omega \in \Lambda_\omega$. Let us assume that the maximum of the jth component Ψ_j of the vector Ψ is attained for

$$q = q_j^*(x, h), \ \xi = \xi_j^*(x, h), \ \omega = \omega_j^*(x, h). \tag{3.10}$$

We have

$$\Psi_j(x, h, q_j^*, \xi_j^*, \omega_j^*) = \max_{q \in \Lambda_q} \max_{\xi \in \Lambda_\xi} \max_{\omega \in \Lambda_\omega} \Psi_j(x, h, q, \xi, \omega). \tag{3.11}$$

We use the notation

$$\Psi_j^*(x, h) \equiv \Psi_j(x, h, q_j^*(x, h), \xi_j^*(x, h), \omega_j^*(x, h)). \tag{3.12}$$

If the maximum of Ψ_j with respect to q, ξ and ω is attained for several different assemblages of functions q, ξ, ω chosen from (3.2) to (3.4), then any collection of these functions can be considered as $\left\{ q_j^*, \xi_j^*, \omega_j^* \right\}$.

Using the introduced notation for Ψ_j^* and the guaranteed approach, we represent the original constraint (3.6) or (3.8) as a system of inequalities

$$\Psi_j^*(x, h) \leq 0. \tag{3.13}$$

Thus the application of the guaranteed approach based on WCS and the fulfillment of the corresponding inequalities enable us, in principal, to transform (analytically or numerically) the original optimal design problem with incomplete information to a conventional optimization problem of minimizing the functional $J(h)$ with respect to h ($h \in H$) under the inequality constraints (3.13). To do this we can use variational techniques and the methods of numerical optimization. In what follows, we consider problems of optimal structural design with incomplete information about loading conditions, damage characteristics or properties of materials, and present some results about transforming these problems to conventional optimization problems.

Note that in this context it is important to develop sensitivity analysis procedures, described in [BH75,HC81,HCK86,BP93,DZ01,Bli46,Ban84,GF63,Kom84,DL88, LL50,Mie99].

3.2 Uncertainties in Loading Conditions

Consider a thin-walled elastic structure loaded by distributed force $q = q(x)$ $\left(x = \{x^1, x^2\}\right)$ and described by a system of differential equations with corresponding boundary conditions written in operator form as

$$A(h)w = q \qquad (3.14)$$

and defined in the domain $\Omega \subset \mathbf{R}^2$ with the boundary $\Gamma = \partial\Omega$. Here $A(h)$ is a linear differential operator with coefficients depending on h and

$$w = w(x) = w(x, h(x)), \ h = h(x), x \in \Omega \qquad (3.15)$$

are respectively displacements and thickness distributions: Ω is the domain, occupied by the mid surface of the thin-walled structure (plate or shell). It is assumed that $h \in H$, where H is a given set of admissible thickness distributions.

The forces in (3.14) are not known *a priori*, but any admissible realization of $q = q(x)$ is supposed to belong to the set Λ_q which contains all admissible loads, i.e.

$$q \in \Lambda_q. \qquad (3.16)$$

If, for example, we optimize a plate of variable thickness and the external normal loads act in one direction and have a total magnitude which does not exceed a given positive constant P_0, then the set Λ_q in (3.16) is given by

$$\Lambda_q = \left\{ q : q(x) \geq 0, \int_\Omega q(x)dx \leq P_0, \ x \in \Omega \right\}. \qquad (3.17)$$

The optimal design problem with incomplete information concerning loading conditions consists in minimizing the volume of material of the thin-walled structure (cost functional)

$$J(h) = \int_{\Omega} h d\Omega \to \min_{h \in H} \qquad (3.18)$$

under the rigidity constraints

$$C_i = C_i(w) \le c_{0i}, \quad i = 1, 2, \ldots, k, \qquad (3.19)$$

where k is a given positive number, $c_{0i} > 0$ are given constants, and C_i are the functionals

$$C_i = \langle w, \chi_i \rangle = \int_{\Omega} w \chi_i d\Omega \qquad (3.20)$$

that characterize the displacements. Here χ_i are given characteristic functions and $\langle \cdot, \cdot \rangle$ is a scalar product. If

$$\chi_i = \delta(x - x_i), \qquad (3.21)$$

where δ is a delta function determined at point $x = x_i$, then

$$C_i = w(x_i) \qquad (3.22)$$

and the constraints (3.19) are imposed only on the values of displacements at given points $x = x_i$.

Thus the minimum of the optimized functional under rigidity constraints is found when we have incomplete information concerning realization of loading conditions. Using the concept of WCS, we consider the worst loads and take them into account in optimization process. We will consider the rigidity constraints in the form

$$\max_{q \in \Lambda_q} C_i(w(x, q)) \le c_{0i}. \qquad (3.23)$$

To do this effectively, we introduce the adjoint variables v_i ($i = 1, 2, \ldots, k$) as solutions of the boundary value problems

$$A(h)v_i = \chi_i, \quad i = 1, 2, \ldots, k. \qquad (3.24)$$

Then taking into account that the behavioral operator A of plates and shells is self-adjoint, i.e. $A^* = A$, we perform the following transformations:

$$C_i = \langle w, \chi_i \rangle = \langle w, A v_i \rangle = \langle A^* w, v_i \rangle = \langle Aw, v_i \rangle = \langle q, v_i \rangle. \qquad (3.25)$$

Thus we will have

$$C_i = \langle q, v_i \rangle \qquad (3.26)$$

and, consequently, as seen from (3.26), the rigidities C_i can be expressed in the form of the functions v_i, which depend explicitly on q and some functions v_i, i.e.

$$C_i = C_i(q, v_i). \tag{3.27}$$

Note that the functions

$$v_i = v_i(x) = v_i(x, h(x)), \ x \in \Omega, \tag{3.28}$$

do not depend on q, so that the functionals C_i are linear functionals with respect to variable q.

Thus we can transform the inequality constraints (3.19) or (3.23), imposed explicitly on the displacement function w and implicitly on the loads q, to the inequality constraints

$$\max_{q \in \Lambda_q} C(q, v_i) = \max_{q \in \Lambda_q} \langle q, v_i \rangle \leq c_{oi}, \ i = 1, 2, \ldots, \kappa \tag{3.29}$$

imposed explicitly on the admissible load q. If the solutions (3.28) of the adjoint boundary value problems (3.24) have been found, then the results of maximization of the linear functionals in (3.29) can be presented as

$$q_{i*} = \arg \left(\max_{q \in \Lambda_q} \langle q, v_i \rangle \right). \tag{3.30}$$

In the particular case that Λ_q is given by Eq. (3.17), we obtain

$$q_{i*} = P_0 \delta (x - x_{i*}), \tag{3.31}$$

$$x_{i*} = \arg \left\{ \max_x v_i(x, h) \right\}. \tag{3.32}$$

After substitution of $q_{i*}(x, h)$ from Eqs. (3.30) into (3.29) we find

$$C_i(h) = C_i \left(q_{i*}(x, h), v_i(x, h) \right) \leq c_{0i}. \tag{3.33}$$

Thus the original optimal design problem with incomplete information (3.14), (3.16), (3.18)–(3.20) is reduced to a conventional optimization problem for the functional (3.18) under the inequality constraints (3.33).

3.3 Some Optimal Solutions

As shown in Section 3.1 the original problem with incomplete information can, in principle, be transformed to a conventional optimization problem with the constraints (3.5), (3.13). In some cases these constraints are expressed as a system of explicit inequalities imposed on the desired design variable $h = h(x)$. In these cases it is convenient to use the following representations for optimal solutions.

Consider the following problem: it is required to find a function $h(x)$ which minimizes the functional

$$J = J(h) = \int_{\Omega} a(x)h(x)d \rightarrow \min_{h} \qquad (3.34)$$

under the constraints

$$h(x) \geq H_i(x), \ i = 1, 2, \ldots, k, \qquad (3.35)$$

where k is a given positive number and $a(z) \geq 0$, $H_i(x) \geq 0$, $i = 1, 2, \ldots, k$, are given functions of $x \in \Omega$, and Ω is a given domain. To minimize the integral (3.34) under the inequalities (3.35), the function $h(x)$ at any point $x \in \Omega$ must be as small as possible but not violate the constraints (3.35). It means that the value h at any point x is determined from the conditions

$$h(x) \rightarrow \min, \ x \in \Omega,$$
$$h(x) \geq H_1(x), \ldots, h(x) \geq H_k(x). \qquad (3.36)$$

The solution of (3.36) and (3.34), (3.35) is given by the formula

$$h_*(x) = \max \{H_1(x), H_2(x), \ldots, H_k(x)\}, \ x \in \Omega. \qquad (3.37)$$

Now consider another case, when the following bilateral inequalities are imposed on the desired design variable $h = h(x)$:

$$H_i^-(x) \leq h(x) \leq H_i^+(x), \ i = 1, 2, \ldots, k, \ x \in \Omega, \qquad (3.38)$$

where $H_i^-(x)$ and $H_i^+(x)$ are given positive functions, κ is a given positive number, Ω is a given domain.

Let us introduce the functions

$$H_{\max}^-(x) = \max \{H_1^-(x), H_2^-(x), \ldots, H_\kappa^-(x)\}, \qquad (3.39)$$
$$H_{\min}^+(x) = \min \{H_1^+(x), H_2^+(x), \ldots, H_\kappa^+(x)\}. \qquad (3.40)$$

If for any $x \in \Omega$, the condition

$$H_{\max}^-(x) \leq H_{\min}^+(x) \qquad (3.41)$$

is satisfied, then the optimal solution $h_*(x)$, minimizing the integral (3.34) under the constraints (3.38), has the form

$$h_* = h_*(x) = H_{\max}^-(x), \ x \in \Omega. \qquad (3.42)$$

There is no optimal solution if the condition (3.41) is violated at some points $x \in \Omega$; in this case there is not only no optimal solution but also no admissible distributions of the design variable.

3.4 Optimal Design with Uncertainties Versus Optimal Multipurpose Design

The wide class of problems known as problems of multipurpose optimization [PS68, Mar70, BK76, Kar79a, Kar79b, KP79, KP80, Kur77, PK77, PK80, EKO90] can be considered as problems with incomplete information. In these problems not only are input values such as forces, damages and material constants characterized by incomplete information, but also the model that determines the behavior of the optimized structure has uncertainties (as an element of the given set of all used models). In particular, we can consider various loading conditions applied to the structural elements.

To explain that the optimization problems with incomplete information are equivalent to certain multipurpose optimal design problems we can consider previously presented problems as problems of optimal beam design, in which we successively subjected the beam to various types of load with the resultant load not exceeding some previously assigned magnitude. The program of loading the beam should include all admissible loads of the given type. An important property of the problems was the following: As we load the beam we permit only bending loads. The tension and torsional loads are not permitted. Therefore, in our optimization problems we always considered the same equations of state (i.e., the beam bending equation). It is at least of equal interest, both theoretically and practically, to study optimum shapes when loads of different types are applied to the beam. In that case, we must consider different defining equations and properties of the structure. As an example we will offer some solutions to problems of this kind, in which we seek the optimum shape of an elastic rod that is subjected consecutively to bending and torsion.

Suppose we wish to minimize the cross-sectional area of an elastic cylindrical bar with torsional rigidity (K) and bending rigidity (C) satisfying the inequalities $K \geq K_0$, $C \geq C_0$, where K_0 and C_0 are given constants. We assume that the bar is not subjected to simultaneous bending and torsion, but that it works either in bending or in torsion as the loads change. We shall offer a rigorous mathematical formulation for this problem. We recall the basic relations concerning the pure bending or pure torsion of bars. Let a cylindrical bar (i.e. one with uniform cross-section) be twisted by a couple M applied to both ends. The resultant angle of twist θ (per unit length of the beam) is proportional to M, i.e., $M = K\theta$, where K is the bar torsional rigidity. To compute the value of the torsional rigidity, we introduce the stress function $\varphi = \varphi(x, y)$, which satisfies the following boundary-value problem:

$$\frac{\partial^2 \varphi}{\partial x^2} + \frac{\partial^2 \varphi}{\partial y^2} = -2, \ (x, y) \in \Omega, \ (\varphi)_\Gamma = 0. \tag{3.43}$$

Here Ω is a simply connected domain occupying the transverse cross section of the bar, and Γ is the boundary of Ω. The torsional rigidity K of the bar is given, in terms of the stress function $\varphi(x, y)$, by the formula

$$K = 2G \iint_\Omega \varphi dx dy, \tag{3.44}$$

where G denotes the shear modulus. We assign the constraint

$$K \geq K_0, \tag{3.45}$$

where K_0 is a positive constant.

Now, let us consider the case in which the rod is used as a beam and is subjected to bending. The most important mechanical property characterizing a beam is its bending rigidity against transverse loading. Assuming that bending takes place in the $y - z$ plane (z is the axis of the beam) and denoting Young's modulus by E, we recall the general formula for bending rigidity

$$C = E \iint_\Omega y^2 dx dy. \tag{3.46}$$

For beams with constant cross section, not only the static rigidity against transverse loading, but also the spectrum of natural frequencies in vibration depends on the magnitude of the rigidity modulus C. By increasing the bending rigidity of a beam (for a given constant mass) we decrease the size of its deflections caused by transverse loads and increase the frequencies of free vibrations. Therefore, constraints on the greatest permissible deflection and on the dynamic properties can be reduced to the following inequality:

$$C \geq C_0, \tag{3.47}$$

where C_0 is a given positive constant. The optimization problem consists of finding the cross-sectional shape of the elastic rod that satisfies conditions of (3.45)–(3.47) and minimizes the area of the cross section

$$S(\Gamma) = \iint_\Omega dx dy \rightarrow \min. \tag{3.48}$$

To formulate this problem, we first analyze all possible types of solutions. To do this, let us first consider the problem of (3.43)–(3.45), and (3.48); i.e. the problem of minimizing the area of transverse cross section with only a single constraint concerning the torsional rigidity. It is easy to show that a dual problem is one of maximizing torsional rigidity for a given area S of rod cross section. In the original problem the optimum rod has a circular cross section [PS51, Pol48] having the radius $r = [2K_0 / (\pi G)]^{1/4}$. The circular cross section is also optimum for the dual problem.

Fig. 3.1 The plane of the design parameters $K \geq 0$, $C \geq 0$

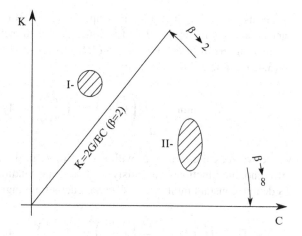

The torsional and bending rigidities of the circular rods are related to each other by the formula $K = 2GC/E = C/(1 + \nu)$ where ν is Poisson's ratio. This functional dependence has a graph that is a straight line separating the plane with parameters K and C as coordinates into two regions I and II, as indicated in Fig. 3.1. The slope of this line varies between 45° and 33.7°, as ν varies between the limits $0 \leq \nu \leq 1/2$. To each rod with a circular cross section corresponds a point on this line in the $K - C$ plane. If constants K_0 and C_0 satisfy the inequality

$$K_0 \geq 2GC_0/E, \tag{3.49}$$

i.e., if points corresponding to such a constant lie in region I (see Fig. 3.1) then the optimum rod has a circular cross section. This follows from the fact that if the inequality of (3.49) is true, then the torsional rigidity of a round rod (which satisfies the inequality of (3.45) is greater than C_0, i.e., satisfies the inequality of (3.47).

If the values of the parameters K_0 and C_0 violate the inequality of (3.49); i.e., they lie in region II of Fig. 3.1, then both conditions of (3.45) and (3.47) must be taken into account in finding the optimum shape of the cross section (which we assume to be convex) of a rod. In this case, we take into account the constraints of (3.45) and (3.47) by the use of Lagrange multipliers. We note that there are no regions in the quadrant $K \geq 0$, $C \geq 0$ in which the optimum shape may be determined by the use of the single inequality (3.47). The explanation for this is that we could let the cross-sectional area approach zero, without violating the convexity condition, while retaining the same value for the bending rigidity of the rod. Let us explain this phenomenon by offering an example of a rectangular cross section with width b and height h, for which $S = bh$ and $C = Ebh^3/12$; we can retain a constant value of bending rigidity $C = C_0$ as $h \to \infty$ while the cross-sectional area S tends to zero.

Let us determine the optimum shape for region II of Fig. 3.1. To obtain necessary conditions for optimality, it is convenient to transform the basic relations by eliminating the differential equation (3.43) from our considerations. The constraint of (3.45) can be written in the form

$$K = \min_{\varphi} \frac{G}{2} \iint_{\Omega} \left(\left(\frac{\partial \varphi}{\partial x} \right)^2 + \left(\frac{\partial \varphi}{\partial y} \right)^2 - 4\varphi \right) dxdy \geq K_0,$$

where we seek the minimum with respect to φ, in the class of once continuously differentiable functions that satisfy the boundary condition $\varphi = 0$ on Γ. Let λ_1 and λ_2 denote constant multipliers. We construct a Lagrangian functional

$$\Pi = \iint_{\Omega} \left[1 + \lambda_1 \frac{G}{2} \left(\left(\frac{\partial \varphi}{\partial x} \right)^2 + \left(\frac{\partial \varphi}{\partial y} \right)^2 - 4\varphi \right) + \lambda_2 E y^2 \right] dxdy,$$

where the last term with the Lagrange multiplier λ_2 corresponds to the constraint (3.46), (3.47). We derive the necessary condition for optimality of the contour Γ:

$$1 - \frac{1}{2} \lambda_1 G \left(\left(\frac{\partial \varphi}{\partial x} \right)^2 + \left(\frac{\partial \varphi}{\partial y} \right)^2 \right) + \lambda_2 E y^2 = 0,$$

which may be rewritten in the form

$$\left(\frac{\partial \varphi}{\partial x} \right)^2 + \left(\frac{\partial \varphi}{\partial y} \right)^2 = \mu_1 + \mu_2 y^2, \tag{3.50}$$

where $\mu_1 = 2/(\lambda_1 G)$ and $\mu_2 = 2\lambda_2 E/(\lambda_1 G)$. To find the constants μ_1 and μ_2, we set $K = K_0$ and $C = C_0$. Thus, the solution of our problem of shape optimization (that is determining the optimum shape of the transverse cross section and of the corresponding stress function $\varphi(x, y)$), may be found by use of (3.43)–(3.47), and (3.50), where strict equality is assumed in (3.45) and (3.47).

Starting with the necessary condition of optimality of (3.50), we look for the solution of our optimization problem, in the form

$$\Gamma : x^2 + ay^2 = b,$$
$$\varphi = N \left(b - x^2 - ay^2 \right), \tag{3.51}$$

where a, b, and N are unknown constants, which must be determined together with the Lagrange multipliers μ_1 and μ_2 by the basic relations of our problem. It is easy to check that for Γ and φ in the form given by (3.51), the boundary conditions of (3.43) is automatically satisfied. Substituting the expression for φ into the optimality condition of (3.50), we obtain

$$x^2 + \left(a^2 - \mu_2/4N\right) y^2 = \mu_1/4N. \tag{3.52}$$

Since (3.51) and (3.52) describe the same curve, we obtain two relations between the unknown coefficients

$$a^2 - \mu_2 \left(4N^2\right) = a \text{ and } \mu_1/\left(4N^2\right) = b.$$

The remaining three equations that are necessary for determination of a, b, μ_1, μ_2, and N we obtain by substituting (3.51) into (3.43) and into the constraints of (3.45) and (3.47). (We use strict equality in (3.45) and (3.47)). Thus, we have the following system of algebraic equations:

$$a^2 - \mu_2/4N^2 = a, \quad \mu_1/4N_2 = b, \quad 1 + a = 1/N,$$
$$b^2 N/a^{1/2} = K_0/\pi G, \quad b^2/a^{3/2} = 4G_0/\pi E.$$

Solving this system for the unknown quantities, we obtain

$$a = 1/(\beta - 1), \quad b = (K_0\beta/\pi G)^{1/2} (1/\beta - 1)^{3/4}, \quad N = (\beta - 1)/\beta,$$
$$\mu_1 = 4 (K_0/\pi G)^{1/2} (\beta - 1)^{5/4}/\beta^{3/2}, \quad \mu = 4 (2 - \beta)/\beta^2,$$

where $\beta = 4C_0G/(EK_0)$. Equation (3.49) implies that $\beta \geq 2$ in the entire region II of Fig. 3.1. For this region, the shape of the cross section for a cylindrical rod of smallest weight, and the corresponding stress function have the form

$$\Gamma : x^2 + \left(\frac{1}{\beta} - 1\right) y^2 = \left(\frac{K_0\beta}{\pi G}\right)^{1/2} (\beta^{-1} - 1)^{3/4},$$
$$\varphi = \left[(K_0\beta/\pi G)^{1/2} (\beta^{-1} - 1)^{3/4} - x^2 - (\beta^{-1} - 1) y^2\right] ((\beta - 1)/\beta). \tag{3.53}$$

Thus, the solution of (3.53) to the optimization problem is fully determined if the material constants E and G and the constants K_0 and C_0 satisfying the constraint $4GC_0/(EK_0) \geq 2$ are given. The cross-sectional area for the optimum rod is

$$S_{opt} = (\pi K_0/G)^{1/2} \left(\beta^2/(\beta - 1)\right)^{1/4}. \tag{3.54}$$

To estimate the effectiveness of the optimization process, we compare S_{opt} with the area of a rod having a circular cross section and having the same bending rigidity C_0 as the optimum rod. For a round rod, we have $S_0 = (4\pi C_0/E)^{1/2}$. The gain attained by optimizing is given by

$$\left(S_0 - S_{opt}\right)/S_0 = 1 - (\beta - 1)^{-1/4}.$$

It is clear from this formula that when $\beta = 2$ (the boundary of the region II), $S_0 = S_{opt}$ and a round rod is optimum. As β increases (with $\beta \geq 2$), the gain according to this formula also increases as the shape is optimized.

In addition to the above problem, we also consider two more basic problems that lead to the optimization condition of (3.50). The first problem consists of maximizing the torsional rigidity of a cylindrical rod ($K \to \max$), with constraints on the area of the cross section and the bending rigidity of the rod

$$C \geq C_0, \quad S \leq S_0. \tag{3.55}$$

In the second problem, we minimize the bending rigidity of the rod ($C \to \max$), with constraints on the area of the cross section and on the torsional rigidity

$$K \geq K_0, \quad S \geq S_0. \tag{3.56}$$

Let us describe the solutions to these problems, without going into details. In the first problem, the first quadrant of the $C - S$ plane ($C \geq 0$, $S \leq 0$) is divided into two regions by the parabola $C = ES^2/(4\pi)$, as shown in Fig. 3.2. If the parameters C_0 and S_0 satisfy the inequality

$$C_0 \leq ES_0^2/4\pi, \tag{3.57}$$

i.e., if they lie in region I of Fig. 3.2, then the cross-sectional area of the rod having the greatest torsional rigidity is a circular disk with radius $r = (\pi S_0)^{1/2}$. In region I, constraints on bending rigidity are automatically satisfied so they have no influence on the optimum shape of the rod.

If the inequality of (3.57) is violated (i.e. the point (C_0, S_0) lies in region II of Fig. 3.2), then both constraints of (3.55) influence the optimum shape of the cross section. As before, the optimum cross section is an elliptic disc and Γ is given by

$$\Gamma : x^2 + ky^2 = \sqrt{k}S_0/\pi, \quad k = (ES_0/4\pi C_0)^2$$

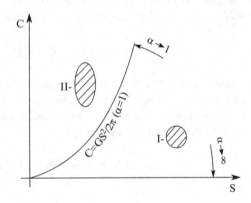

Fig. 3.2 The plane of the design parameters $C \geq 0$, $S \geq 0$

Fig. 3.3 The plane of the
design parameters $K \geq 0$,
$S \geq 0$

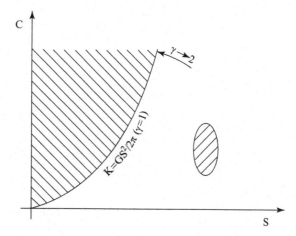

with $\varphi = \left[\sqrt{k} S_0 / \left(\pi - x^2 - ky^2 \right) \right] / (1 + k)$. The torsional rigidity is equal to

$$K_{\mathrm{opt}} = \sqrt{\kappa} G S_0^2 / \pi \, (1 + \kappa).$$

Now let us describe the solution of the second problem. As in the preceding problem, in the case of (3.56), the first quadrant $K \geq 0$, $S \geq 0$ is divided into two regions by the parabola $K = GS^2/2\pi$, shown in Fig. 3.3.

One aspect of this problem is completely different from the preceding ones. If the parameters K_0 and S_0 are such that $K_0 \geq GS_0^2/2\pi$; i.e., if the point (K_0, S_0) lies in the shaded region shown in Fig. 3.3, this problem has no solutions. We offer the following explanation: For an arbitrary cylindrical rod having the cross sectional area $S \leq S_0$, the torsional rigidity satisfies the inequality $K \leq GS^2/2\pi$. In the unshaded region of Fig. 3.3 in which $K_0 \leq GS_0^2/2\pi$, the optimum shape of the boundary and the corresponding stress function φ satisfy

$$x^2 + \delta y^2 = \sqrt{\delta} S_0 / \pi, \quad \varphi = \left[\sqrt{\delta} S_0 / (\pi - x^2 - \delta y^2) \right] (1 + \delta)$$

and the corresponding bending rigidity is given by

$$K_0 \geq GS_0^2/2\pi.$$

We comment that the technique explained in this section, consisting of analyzing all possible types of solutions, may be directly generalized to multipurpose design problems with a large number of constraints.

Chapter 4
Optimal Design of Beams and Plates with Uncertainties

4.1 Beams Having the Smallest Weight with Constraints on Strength

The equilibrium state of a simply supported beam of length l, situated in the $x - y$ plane, with its axis lying on the x-axis and loaded by exterior distributed load $q(x)$ parallel to the y axis, is given by the following system of equations and boundary conditions

$$\frac{dM}{dx} = Q, \quad \frac{dQ}{dx} = -q, \tag{4.1}$$

$$M(0) = M(l) = 0, \tag{4.2}$$

where $M = M(x)$ and $Q = Q(x)$ denote, respectively, bending moment and shear load acting on the cross section of the beam, perpendicular to the x-axis. The beam has a rectangular cross-sectional area of width b and height h.

We assume that the load applied to the beam is positive (the direction of load action coincides with the positive direction of the y-axis) and the resultant force does not exceed a given magnitude P_0, i.e.

$$q(x) \geq 0, \quad \int_0^l q(x)dx \leq P_0. \tag{4.3}$$

The stresses σ_x and τ_{xy} are calculated according to the formulas

$$\sigma_x = \frac{M\zeta}{I}, \quad \tau_{xy} = \frac{1}{b}\frac{\partial}{\partial x}\left(\frac{Md}{I}\right), \tag{4.4}$$

$$I = \frac{bh^3}{12}, \quad d = \frac{b}{2}\left(\frac{h^2}{4} - \zeta^2\right).$$

The coordinate ζ is measured from the centroid of the beam cross-sectional area and varies between the limits $-h/2 \leq \zeta \leq h/2$.

N.V. Banichuk and P.J. Neittaanmäki, *Structural Optimization with Uncertainties*, Solid Mechanics and Its Applications 162, DOI 10.1007/978-90-481-2518-0_1, © Springer Science+Business Media B.V. 2010

For an arbitrary load satisfying (4.3), the normal and tangential stresses σ_x and τ_{xy} must satisfy the strength conditions

$$\psi_1 \equiv |\sigma_x| - \sigma_0 \leq 0, \quad \psi_2 \equiv |\tau_{xy}| - \tau_0 \leq 0, \tag{4.5}$$

where σ_0 and τ_0 are given constants.

The conditions (4.1)–(4.5) are fully symmetric with respect to the point $x = l/2$. We shall, therefore, consider the range $l/2 \leq x \leq l$. Integrating Eq. (4.1) with the indicated boundary conditions (4.2), we derive the following formulas for M and Q:

$$M(x) = \int_0^l K(x,t)q(t)dt, \tag{4.6}$$

$$Q(x) = \int_0^l T(x,t)q(t)dt, \tag{4.7}$$

where

$$K(x,t) = t\left(1 - \frac{x}{l}\right), 0 \leq t \leq x, \tag{4.8}$$

$$K(x,t) = x\left(1 - \frac{t}{l}\right), x \leq t \leq l, \tag{4.9}$$

$$T(x,t) = -\frac{t}{l}, 0 \leq t \leq x, \tag{4.10}$$

$$T(x,t) = 1 - \frac{t}{l}, x < t \leq l. \tag{4.11}$$

Since the functions $K(x,t)$ and $q(x)$ are positive, it follows that $M(x) \geq 0$. We fix a point $x \in [l/2, l]$ and consider the set of values that can be attained by $M(x)$ and $Q(x)$ for all possible admissible loads $q = q(x)$ satisfying (4.3). We denote by $\max_q M(x)$ and $\max_q |Q(x)|$, respectively, the maximum values of the bending moment and of the shear force and we claim that, for $l/2 \leq x \leq l$,

$$\max_q M(x) = P_0 x\left(1 - \frac{x}{l}\right), \quad \max_q |Q(x)| = \frac{P_0 x}{l}. \tag{4.12}$$

To prove this assertion, we use expressions (4.8)–(4.11) and make the following estimates:

$$M(x) \leq \max_{t \in [0,l]} K(x,t) \int_0^l q(t)dt \leq P_0 x\left(1 - \frac{x}{l}\right). \tag{4.13}$$

We also observe that for the admissible load $q(t) = P_0\delta(t - x)$, the value of the bending moments is $M(x) = P_0x(l - x)/l$, where δ denotes the Dirac delta function. Combining this with the estimates given above, we see that the assertion (4.12) is true for $M(x)$.

To prove the validity of Eq. 4.12 for Q, we carry out analogous estimates,

$$|Q(x)| \leq \text{vrai} \max_t |T(x,t)| \int_0^l q(t)\, dt \leq \frac{xP_0}{l}, \qquad (4.14)$$

where $\text{vrai} \max_t |T|$ denotes the essential maximum with respect to t ($0 \leq t \leq l$) of the piecewise continuous function $T(x,t)$, which has a discontinuity at $t = x$. We substitute the admissible load $q(t) = P_0\delta(t - x^1)$ with $l/2 \leq x^1 \leq x$ into (4.7) for the shear load Q and compute the corresponding integral. We obtain

$$|Q(x)| = \frac{P_0x^1}{l}.$$

In the limit as $x^1 \to x - 0$ we have

$$\lim_{x^1 \to x-0} |Q(x)| = \frac{P_0x}{l}.$$

Taking into account the inequality

$$|Q(x)| \leq \frac{P_0x}{l},$$

we obtain the expression (4.12) for Q.

We utilize the derived properties of the functions $M(x)$ and $Q(x)$ to find explicit formulas Ψ_1^* and Ψ_2^* (also see Section 3.1). First, let us find formulas for Ψ_1 and Ψ_2. Making use of (4.4)–(4.12), we have

$$\Psi_1 \equiv \frac{12}{bh^3} \int_0^l K(x,t)q(t)dt - \sigma_0, \qquad (4.15)$$

$$\Psi_2 \equiv \frac{6}{b} \left| \left(\frac{1}{4h} - \frac{\zeta^2}{h^3}\right) \int_0^l T(x,t)q(t)dt + \left(\frac{3\zeta^2}{h^4} - \frac{1}{4h^2}\right)\frac{dh}{dx} \int_0^l K(x,t)q(t)dt \right| - \tau_0.$$

$$(4.16)$$

We compute the maximum of the function Ψ_1, with respect to q using (4.3) and with respect to ζ, on the interval $-h/2 \leq l \leq h/2$, by using the estimate (4.12) for $\max M$

$$\Psi_1^* = \frac{6P_0x}{bh^2}\left(1 - \frac{x}{l}\right) - \sigma_0. \qquad (4.17)$$

We shall also determine the maximum of the function Ψ_2 with respect to q and ζ.

We note that the expression inside the absolute value signs in the formula for Ψ_2 is a linear function of ζ^2. Consequently, the maximum of Ψ_2, regarded as a function of ζ^2, is attained on the interval $-h/2 \leq \zeta \leq h/2$, either when $\zeta^2 = h^2/4$ or else when $\zeta^2 = 0$. Following this observation, we have

$$\Psi_2^* = \max(\chi_1, \chi_2) - \tau_0, \tag{4.18}$$

where

$$\chi_2 = \frac{3P_0 x}{bh^2} \left(1 - \frac{x}{l}\right) \left|\frac{dh}{dx}\right|, \tag{4.19}$$

$$\chi_1 = \max\left(\frac{3P_0}{2bh} \left|\frac{x}{l} + \frac{x}{h}\left(1 - \frac{x}{l}\right)\frac{dh}{dx}\right|, \quad \frac{3P_0}{2bh}\left|\frac{x}{l} - 1 + \frac{x}{h}\left(1 - \frac{x}{l}\right)\frac{dh}{dx}\right|\right). \tag{4.20}$$

Similar, but more detailed, arguments were offered in [Ban75, Ban76, Ban83].

Consider at first the beam having a rectangular cross-sectional area of constant height h and variable width $b = b(x)$. The function $b = b(h)$ determining the shape of the beam is the unknown function. The problem of optimizing the beam shape consists of finding a function $b = b(x)$ that, for any arbitrary load $q = q(x)$ satisfying conditions (4.3), satisfies the constraints (4.5) and minimizes the integral

$$J = J(b) = \gamma h \int_0^l b(x)dx \to \min_b, \tag{4.21}$$

i.e. the weight of the beam, where γ denotes the specific weight of the material.

In the considered case $dh/dx = 0$ and the inequalities (3.13), written with the help of the expression (4.17)–(4.20), are reduced to the following explicit constraints:

$$b(x) \geq \frac{6P_0 x}{\sigma_0 h^2} \left(1 - \frac{x}{l}\right), \tag{4.22}$$

$$b(x) \geq \frac{3P_0 x}{2hl\tau_0}, \quad \frac{l}{2} \leq x \leq l, \tag{4.23}$$

and solution of the optimization problem (4.21)–(4.23) is given by the formula

$$b_*(x) = \max\left\{\frac{6P_0 x}{\sigma_0 h^2}\left(1 - \frac{x}{l}\right), \frac{3P_0 x}{2hl\tau_0}\right\}, \quad \frac{l}{2} \leq x \leq l. \tag{4.24}$$

Optimal width distribution can be also written as (see also Fig. 4.1)

Fig. 4.1 Optimum width distribution for a one-half of the beam

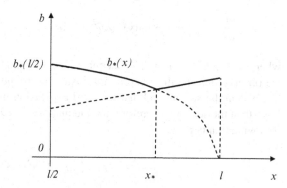

$$b_*(x) = \begin{cases} \frac{6P_0x}{\sigma_0 h^2}\left(1 - \frac{x}{l}\right), & \frac{l}{2} \leq x \leq x_*, \\ \frac{3P_0x}{2hl\tau_0}, & x_* \leq x \leq l, \end{cases} \qquad (4.25)$$

where the value x_* is found with the help of the continuity condition, and is $x_* = l - h\sigma_0/4\tau_0$. If $h\sigma_0/2\tau_0 \geq l$ then the optimal shape of the beam has the form

$$b_*(x) = \frac{3P_0x}{2hl\tau_0}, \qquad \frac{l}{2} \leq x \leq l. \qquad (4.26)$$

Consider now the beam having a rectangular cross-section of constant width and variable height $h = h(x)$. The problem of optimizing the beam shape consists of finding a function $h = h(x)$ that for any arbitrary load $q = q(x)$, satisfying conditions (4.3), satisfies the constraints (4.5) and minimizes the integral

$$J = J(h) = 2\gamma b \int_{l/2}^{l} h(x)dx \rightarrow \min_h. \qquad (4.27)$$

We apply the expressions (4.17)–(4.20) for functions Ψ_1^* and Ψ_2^* derived above and formulate conditions that the function $h = h(x)$ must satisfy, if the inequalities (4.5) are to be satisfied. Substituting expression (4.17) for Ψ_1^* into the first inequality (3.13), we obtain the condition

$$h(x) \geq \Phi_1(x) \equiv \left[\frac{6P_0x}{b\sigma_0}\left(1 - \frac{x}{l}\right)\right]^{1/2}. \qquad (4.28)$$

Substituting the formula for Ψ_2^* given in (4.18)–(4.20) into the second inequality (3.13), we will have

$$\frac{dh}{dx} \leq \frac{1}{x(l-x)} \min\left(\lambda h^2, 2\lambda h^2 - xh, 2\lambda h^2 + (l-x)h\right), \qquad (4.29)$$

$$\frac{dh}{dx} \geq \frac{1}{x\,(l-x)} \max\left(-\lambda h^2, -2\lambda h^2 - xh, -2\lambda h^2 + (l-x)\,h\right), \qquad (4.30)$$

where $\lambda = bl\tau_0/(3P_0)$. The inequalities (4.29) and (4.30) may be simplified if we observe that the third expression inside the round brackets in (4.29) is always larger than the second and that the third expression inside the brackets in (4.30) is large than the second. Therefore, the inequalities (4.29), (4.30) can be written in the following manner:

$$\frac{dh}{dx} \leq \frac{1}{x\,(l-x)} \min\left(\lambda h^2, 2\lambda h^2 - xh\right), \qquad (4.31)$$

$$\frac{dh}{dx} \geq \frac{1}{x\,(l-x)} \max\left(-\lambda h^2, -2\lambda h^2 + (l-x)\,h\right). \qquad (4.32)$$

Now, let us decompose the domain

$$O\left(\frac{l}{2} \leq x \leq l,\ h \geq 0\right) \qquad (4.33)$$

in which we seek the solution of our optimization problem, into three sub-domains

$$O_1\left(\frac{l}{2} \leq x \leq l,\ h \geq \frac{x}{\lambda}\right), \qquad (4.34)$$

$$O_2\left(\frac{l}{2} \leq x \leq l,\ \frac{l-x}{\lambda} \leq h \leq \frac{x}{\lambda}\right), \qquad (4.35)$$

$$O_3\left(\frac{l}{2} \leq x \leq l,\ 0 \leq h \leq \frac{l-x}{\lambda}\right). \qquad (4.36)$$

In each of these subregions, the inequalities (4.31) and (4.32) assume the form

$$-\frac{\lambda h^2}{x\,(l-x)} \leq \frac{dh}{dx} \leq \frac{\lambda h^2}{x}\,(l-x),\quad (x,h) \in O_1, \qquad (4.37)$$

$$-\frac{\lambda h^2}{x\,(l-x)} \leq \frac{dh}{dx} \leq \frac{2\lambda h^2 - xh}{x\,(l-x)},\quad (x,h) \in O_2, \qquad (4.38)$$

$$-\frac{2\lambda h^2 - (l-x)\,h}{x\,(l-x)} \leq \frac{dh}{dx} \leq \frac{2\lambda h^2 - xh}{x\,(l-x)},\quad (x,h) \in O_3. \qquad (4.39)$$

For any $(x,h) \in O_1$, the inequalities given by (4.37) cannot be unsolvable. For the inequalities (4.38) to be solvable in the region O_2 and for the inequalities (4.39) to be solvable in the region O_3, the following conditions must be satisfied:

$$h \geq \frac{x}{3\lambda}, \; h \geq \frac{l}{4\lambda}. \tag{4.40}$$

Thus, the original problem of optimizing the shape of a rectangular beam with variable thickness has been reduced to finding a continuous function $h(x)$, satisfying the constraints (4.28) and (4.40) and the differential inequalities (4.37)–(4.39), such that $h(x)$ minimizes the integral (4.27).

The functions $h(x)$ satisfying the inequalities (4.28) and (4.37)–(4.39) are called admissible. Some properties of admissible functions will be derived and applied to future studies. Let us consider a function $h = h(x)$ that passes through the point

$$\left(x^0, h^0\right) \in O_2 + O_3, \; h^0 = h(x^0).$$

It is a consequence of the inequalities (4.38) and (4.39) that

$$h(x) \leq v(x)$$

for all $x > x^0$, where $v(x)$ stands for a solution of the differential equation

$$\frac{dv}{dx} = \frac{2\lambda v^2 - xv}{x \, (l - x)} \tag{4.41}$$

satisfying the initial condition

$$v\left(x^0\right) = h^0. \tag{4.42}$$

Integrating the differential equation (4.41), making use of the initial condition (4.42), we obtain

$$v(x) = \frac{l - x}{2\lambda \ln\left(\frac{c}{x}\right)}, \tag{4.43}$$

$$c = x^0 \exp\left[\frac{l - x^0}{2\lambda h^0}\right]. \tag{4.44}$$

Let us consider the behavior of the integral curves $v(x)$ depending on the position of the initial point $\left(x^0, h^0\right)$. If the values $\left(x^0, h^0\right)$ are such that $c > l$, then it is easy to see from the expressions (4.43), (4.44) that $v(x) \to 0$ as $x \to l$ (see Fig. 4.2). For initial conditions $\left(x^0, h^0\right)$ such that $c < l$, the integral curves approach infinity as x approaches c.

However, if the magnitudes of x^0 and h^0 satisfy the relation

$$x^0 \exp\left[\frac{l - x^0}{2\lambda h^0}\right] = l, \tag{4.45}$$

then as $x \to l$ both the numerator and denominator in the expression (4.43) for v approach zero.

Fig. 4.2 Behavior of the
auxiliary function $v(x)$

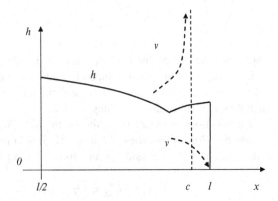

Applying L'Hospital's rule, we have in the limit

$$v(l) = \frac{l}{2\lambda}. \tag{4.46}$$

For $c = l$, we denote the function $v(x)$ by $\Phi_2(x)$.

Using these properties of the function $v(x)$, we can show that the solution of our problem assumes the form (see Fig. 4.2)

$$h_*(x) = \begin{cases} \Phi_1(x), & \frac{l}{2} \le x \le x_*, \\ \Phi_2(x), & x_* \le x \le l, \end{cases} \tag{4.47}$$

$$\Phi_1(x) \equiv \left[\frac{6P_0 x}{\sigma_0 b} \left(1 - \frac{x}{l} \right) \right]^{1/2}, \quad \Phi_2(x) \equiv \frac{l - x}{2\lambda \ln \left(\frac{l}{x} \right)} \tag{4.48}$$

or formulating it differently

$$h_*(x) = \max \{ \Phi_1(x), \Phi_2(x) \}, \frac{l}{2} \le x \le l. \tag{4.49}$$

The value x_* may be determined as follows: If the parameters of our problem are such that $\Phi_2(x) \ge \Phi_1(x)$ everywhere on the interval $\frac{l}{2} \le x \le l$, then $h_*(x) = \Phi_2(x)$ and x_* in (4.47) may be taken as $x_* = l/2$. It is easy to check that this is true for

$$v \equiv \frac{6P_0 \sigma_0}{lb\tau_0^2} \ge 16 (\ln 2)^2. \tag{4.50}$$

In a different situation, when $v < 16 (\ln 2)^2$ there may exist a subinterval of $l/2 \le x \le l$ in which $\Phi_2(x) < \Phi_1(x)$ and another subinterval in which $\Phi_2(x) > \Phi_1(x)$. In this case, the optimum design consists of two parts, determined by our formula, and the point x_* may be found from the equality $\Phi_1(x_*) = \Phi_2(x_*)$ which may be restated as

$$\left[\frac{x_*}{\nu\,(l - x_*)}\right]^{1/2} \ln\left(\frac{l}{x_*}\right) = \frac{1}{4}. \tag{4.51}$$

The proof of optimality for a solution (4.47)–(4.51) consists of showing that the function $h_*(x)$ is admissible, i.e. satisfies the condition (4.28) and (4.37)–(4.40) and that no other admissible function $h(x)$ exists for which the functional J has a smaller value than for $h_*(x)$ given by (4.47)–(4.51).

Let us first consider the case $\nu < 16 \ln^2 2$ and determine the region in which x can vary so that the inequalities (4.37)–(4.40) are true if we substitute $h = \Phi_1(x)$. We carry out some elementary substitutions and conclude that the inequality (4.37) is satisfied if

$$\frac{l}{2} \le x \le \beta_1 \equiv \frac{l}{2}\left\{1 + \frac{2}{\sqrt{4 + \nu}}\right\}, \tag{4.52}$$

the inequality (4.38) is satisfied if

$$\frac{l}{2} \le x \le \beta_2 \equiv \frac{l}{2}\left\{1 + \frac{1}{4}\sqrt{16 - \nu}\right\} \tag{4.53}$$

and the inequality (4.39) is satisfied if

$$\frac{l}{2} \le x \le \beta \equiv \min\left(\beta_1, \beta_2\right). \tag{4.54}$$

It is easy to check that

$$\frac{x}{2\lambda} < \Phi_2(x) < \frac{x}{\lambda}$$

if $l/2 \le x < l$, and that $\Phi_2(x) = x/(2\lambda)$ if $x = l$. In particular, it follows that $x_1 = 16l/(\nu + 16)$ is a root of the equation

$$\frac{x_1}{2\lambda} = \Phi_1(x_1)$$

that satisfies the inequality $x_1 \ge x_*$. Therefore to prove the inequality $\beta \ge x_*$, it suffices to confirm the inequality $\beta \ge x_1$ for $0 \le \nu \le 16 \ln^2 2$, which can be done by elementary manipulations. The curve $\Phi_2(x)$ defined in (4.48) lies either in the region O_2 or in $O_2 + O_3$ ($\Phi_2(x) \le x/\lambda$), depending on the value of the parameter ν. The inequalities (4.38) and (4.39) are satisfied, since for $h = \Phi_2(h)$ the right-hand side inequalities in (4.38) and (4.39) become exact equalities. This follows from our construction of the function $\Phi_2(x)$. Consequently, the function $h(x)$ given by expressions (4.47)–(4.51) is admissible if $0 \le \nu \le 16 \ln^2 2$. For $\nu \ge 16 \ln^2 2$, the function $\Phi_2(x)$ ($l/2 \le x \le l$) is also admissible, because for $h = \Phi_2(x)$ the conditions (4.38) and (4.39) are satisfied.

We shall now prove that the admissible function $h(x)$ given by (4.47)–(4.51) is optimum. It suffices to show that the graph of any other admissible function $h(x)$ does not lie below the curves $\Phi_1(x)$ and $\Phi_2(x)$. For $l/2 \le x \le x_*$ the admissible functions $h(x)$ does not lie below the graph of $\Phi_1(x)$, as can be seen from

(4.28), i.e. $h(x) \geq \Phi_1(x)$. We shall prove that for $x_* \leq x \leq l$ the inequality $h(x) \geq \Phi_2(x)$ is true. Let us assume the contrary, i.e., we assume that at some point $x = x'$ $(x_* \leq x' \leq l)$ an admissible function $h(x)$ satisfies the inequality $h(x') < \Phi_2(x')$. However, as we have already indicated for any admissible function $h(x)$ passing through the point $[x', h(x')]$, we have the limit $h(x) \rightarrow 0$ as $x \rightarrow l$. Hence the trajectory $h(x)$ originating at the point $[x', h(x')]$ must enter the forbidden region defined by the inequality (4.40), and violate the hypothesis that the function $h(x)$ is admissible. This contradiction proves the inequality $h(x) \geq \Phi_2(x)$ for $x_* \leq x \leq l$. Consequently, (4.47)–(4.51) provides a solution to our optimization problem.

For the optimum solution, we estimate the magnitude of the stresses σ_x and τ_{xy} that result from application of a concentrated (point) load P_0. Let us denote by σ_x the maximum value of the normal stress acting on the perpendicular cross section (i.e., along the ζ axis), while τ_{xy} is the maximum value of the shear stresses. If the point load P_0 is applied to the optimum beam at the point ξ ($l/2 \leq \xi \leq x_*$), the magnitudes of $\sigma_x(x)$ and $\tau_{xy}(x)$ satisfy the inequalities $\sigma_x(x) \leq \sigma_0$, $\tau_{xy}(x) \leq \tau_0$ ($l/2 \leq x \leq l$), as a consequence of (4.4)–(4.11) and (4.47)–(4.51). The equality $\sigma_x = \sigma_0$ is satisfied when $x = \xi$. If $x_* \leq \xi \leq l$, then $\sigma_x \leq \sigma_0$ and $\tau_{xy} \leq \tau_0$. In this case, the upper bound value of $\tau_{xy} = \tau_0$ for the shear stress is attained at $x = \xi$ and the equality $\sigma_x = \sigma_0$ takes place when $x = \xi = x_*$. Thus, when a point load is applied to the optimum beam, at an arbitrary point x in the interval $[l/2, l]$ the upper bound stress condition is attained only at that point. Consequently, when we design a beam for a fixed total load, we have additional ways of optimization. This has been confirmed in cases for which such optimum shapes were found (we shall not present the details here). Consequently, for problems considered here (in the general sense indicated above), there is no "worst" load and the optimum beam design for the class of admissible loads (4.3) is not optimum for any specific load distribution that belongs to that class.

4.2 Some Rigidity Optimization Problems for Elastic Beams and Plates

In Section 4.1 we considered the problems of optimizing the shape of a beam having least weight (or volume), with constraints on its strength. It is interesting to consider some analogous problems with constraints on the maximum deflection. These problems are duals of the problem of minimizing the maximum deflection, with a given weight or total volume. A solution of each of these problems may be obtained from the other by means of a simple computation. In the following discussion we present the research results published in [Ban75, Ban76, Ban83]. We consider the problem of minimizing the maximum deflection, assuming as before that the applied load belongs to a class defined by the inequalities (4.3).

4.2.1 Minimization of Elastic Deflection

The deflection function $w(x)$ of an elastic beam obeys the differential equation

$$\frac{d^2}{dx^2}\left(EI\frac{d^2w}{dx^2}\right) = q \tag{4.55}$$

with some boundary conditions at the points $x = 0$ and $x = l$. The beam length l and volume V are given. The function $S(x)$ (area of the cross section) is related to l and V by the equality

$$\int_0^l S(x)dx = V. \tag{4.56}$$

In the problem treated here, the cross-sectional area $S(x)$ is the unknown function, and the constraint (4.56) is regarded as an isoperimetric condition assigned to $S(x)$. The function $S(x)$ is related to the moment of inertia $I(x)$ by the equality $EI(x) = C_\alpha S^\alpha(x)$, where α can assume either one of the values $\alpha = 1, 2, 3$, and C_α are given constants depending on the type of the beam cross-section (see, for example, [Ban90]).

Let us consider $S(x)$ as given, and determine the number J that represents the absolute value of the greatest deflection attained, for any admissible load from (4.3), i.e.

$$J = \max_q \max_x w. \tag{4.57}$$

The maximum with respect to x can be computed for all x in the interval $0 \le x \le l$, while the maximum with respect to q is taken over all possible admissible loads, i.e., loads satisfying (4.3). The number J is a functional defined over a class of functions satisfying the equality (4.56).

Let us now formulate the optimization problem. Among all admissible functions $S = S(x)$ defining the shape of the beam and satisfying equality (4.56), we wish to find the one that minimizes J, i.e.

$$J_* = \min_S J = \min_S \max_q \max_x w. \tag{4.58}$$

In other words, we need to find the optimum shape of a beam having the smallest maximum deflection. Below we determine the optimum shape, using the minimax approach. Let us integrate Eq. (4.55) with simply supported boundary conditions. We have

$$w = \int_0^l \Phi(\xi, x) q(\xi) \, d\xi, \tag{4.59}$$

$$\Phi(\xi, x) = \begin{cases} \chi(\xi, x), & 0 < \xi \le x, \\ \chi(x, \xi), & x \le \xi < l, \end{cases}$$

$$\chi(\xi, x) \equiv \left(1 - \frac{x}{l}\right)\left(1 - \frac{\xi}{l}\right) \int_0^\xi \frac{t^2 \, dt}{EI(t)} + \xi \left(1 - \frac{x}{l}\right) \int_\xi^x \frac{(1 - t/l) \, dt}{EI(t)}$$

$$+ \xi x \int_x^l \frac{(1 - t/l)^2}{EI(t)} \, dt.$$

Using the expressions (4.59) and the constraints (4.3), it is easy to show that

$$w \le P_0 \max_\xi \Phi(\xi, x) = P_0 \Phi(c, x)$$

for any admissible $q(\xi)$ satisfying (4.3), and that

$$w = P_0 \Phi(c, x), \tag{4.60}$$

when $q(\xi) = P_0 \delta(\xi - c)$. Here c denotes the coordinates of the point where $\Phi(\xi, x)$ attains its maximum with respect to ξ, with $0 < c < l$. It follows from our estimates that the maximum deflection is attained at the point $\xi = c$. Because of this property in solving our problem, we need only consider the effects of concentrated loads on the deflection of the beam.

The scheme for solving the optimization problem consists of the following steps. First let us determine the function $S_0(x)$ that minimizes the deflection w at the point $x = l/2$ when a concentrated load is applied at this point. The magnitude of this deflection is denoted by w_0. We shall prove later in this section that for $S = S_0(x)$, when the point load P_0 is applied at any point ξ in the interval $0 < \xi < l$, the magnitude of any deflection satisfies the inequality $w \le w_0$.

Because of this property of the function $S_0(x)$, it is easy to conclude that the $S_0(x)$ solves our original problem and that $w_0 = J_*$. To complete this argument let us assume that there exist another design $S_1(x)$ for which $J < w_0$. Then the deflection w (at the point $x = l/2$ with a load P_0 applied at $\xi = l/2$) corresponding to the design $S_1(x)$, satisfies the inequality

$$w \le J \le w_0.$$

But this is a contradiction, because $S_0(x)$ was the design that minimized the deflection w at the point $x = l/2$. Consequently, $S_0(x)$ solves our original problem (4.55)–(4.58) and (4.3). Finding the function $S_0(x)$ is reduced to the solution of an isoperimetric variational problem of minimizing the deflection $w(l/2)$ for all admissible choices of $S(x)$, under the condition (4.56). Applying the Lagrange multiplier technique and performing some elementary substitutions, we derive

$$S_0(x) = \frac{(\alpha + 3)V}{(\alpha + 1)l} \left[\frac{16g(x)}{l^2} \right]^{1/(\alpha+1)}, \tag{4.61}$$

$$w_0 = \frac{P_0 V l^2}{16 C_\alpha} \left[\frac{(\alpha + 1)l}{(\alpha + 3)V} \right],$$

where

$$g(x) = \frac{x^2}{4} \quad \text{for } 0 \le x \le \frac{l}{2}, \quad g(x) = \frac{l(l - x)}{4} \quad \text{for } \frac{l}{2} \le x \le l.$$

Later we shall prove that the deflection w, for the shape $S_0(x)$ determined above, occurs as a result of application of the load $q = P_0 \delta(t - \xi)$ and that it satisfies the inequality $w \le w_0$, for any x and ξ in the interval $(0, l)$.

We consider the case $\alpha = 1$ and find, for the shape of S_0, the distribution due to the action of a point load, in accordance with (4.59) and (4.61). To shorten our arguments, we can apply Betti's reciprocal theorem: For fixed values of x and ξ, if we apply a unit point load at point ξ and compute the deflection at point x, this is equal to the deflection computed at ξ if the unit load is applied at point x. Therefore, we have

$$w = \omega_1(\xi, x), \ 0 < \xi \le x \le l/2,$$
$$w = \omega_1(x, \xi), \ 0 < x \le \xi \le l/2,$$
$$w = \omega_2(\xi, x), \ 0 < \xi \le l/2 \le x < l, \tag{4.62}$$
$$w = \omega_2(x, \xi), \ 0 < x \le l/2 \le \xi < l,$$

$$\omega_1(\xi, x) \equiv \frac{P_0 l^2 \xi}{8 C_1 v} \left[\frac{\xi x}{l} - \xi + \frac{x}{2} + \frac{x^2}{l} - 2x \ln \left(\frac{2x}{l} \right) \right],$$

$$\omega_2(\xi, x) \equiv \frac{P_0 l^2 \xi}{8 C_1 v} \left[\xi \left(\frac{x}{l} - 1 \right) + \frac{l + x}{2} - \frac{x^2}{l} \right].$$

Let us substitute the expressions given in (4.61) with $\alpha = 1$ into the inequality $w \le w_0$, and use (4.62). We also make use of the symmetry that is apparent in (4.62) to reduce the number of inequalities from four to two. We introduce the dimensionless variables $x' = x/l$ and $\xi' = \xi/l$, and omit the primes. Our inequalities now become

$$(1-x)\,\xi^2 + \xi x \left(\frac{1}{2} + x - 2\ln 2x\right) + \frac{1}{8} \geq 0, \quad 0 < \xi \leq x < \frac{1}{2},$$

$$(1-x)\,\xi^2 + \xi \left(\frac{1}{2}(1+x) - x^2\right) + \frac{1}{8} \geq 0, \quad 0 < \xi \leq \frac{1}{2} \leq x < 1. \quad (4.63)$$

Let us look at the first inequality in (4.63). The expression given in the left hand side of this inequality is a quadratic expression in ξ, containing three terms, with a positive coefficient of ξ^2. This expression will be nonnegative for $0 < \xi \leq x \leq 1/2$ if its discriminant is nonpositive, i.e., if

$$2x^2 \left(\frac{1}{2} + x - 2\ln 2x\right)^2 - 1 + x \leq 0.$$

This inequality may be transformed into

$$T_1(x) \leq T_2(x),$$

$$T_1(x) \equiv \sqrt{2}\left(\frac{1}{2} + x - 2\ln 2x\right), \qquad (4.64)$$

$$T_2(x) \equiv \frac{1}{x}\left(\sqrt{1-x}\right).$$

At the point $x = 1/2$ the function $T_1(x)$ and $T_2(x)$ are equal, i.e., $T_1(1/2) = T_2(1/2) = \sqrt{2}$. As $x \to 0$, we have $T_1 \to \infty$ and $T_2 \to \infty$. Hence, to prove (4.64), it suffices to show that the derivatives dT_1/dx and dT_2/dx satisfy the inequality

$$\frac{dT_1}{dx} \geq \frac{dT_2}{dx}, \quad 0 \leq x \leq \frac{1}{2}.$$

Substituting the derivatives

$$\frac{dT_1}{dx} = \sqrt{2}\left[1 - \frac{2}{x}\right],$$

$$\frac{dT_2}{dx} = \frac{x-2}{2x^2\sqrt{1-x}}$$

into these inequalities and performing some elementary manipulations, we arrive at the inequality

$$8x^2\,(1-x) \leq 1$$

that must be satisfied for all x in the interval of definition $0 \leq x \leq 1/2$. Hence, the proof of the first inequality (4.63) is complete.

We shall now prove the second inequality in (4.63). As in the preceding case, the expression in the left-hand side of this inequality is quadratic in ξ, having three terms, and the coefficient of ξ^2 is positive. In the interval $1/2 \leq x \leq 1$ the discriminate of this polynomial is nonpositive (this can be shown easily). Consequently, the second inequality in (4.63) is true. Hence, in the case $\alpha = 1$, the solution of the original optimization problem (4.55)–(4.58) is the function $S = S_0(x)$ which is given by (4.61).

Note that the cases $\alpha = 2$ and $\alpha = 3$ are considered in [Ban83].

4.2.2 Rigidity Estimation for Thin-Walled Elastic Structures and Worst Case Loading

Let us describe another technique for estimating the rigidity of thin-walled structures (plates or shells), which reduces to a computation of displacements that are caused by concentrated loads (see [Ban83, Ban90]). Let w denote the displacement function for a thin-walled structure obtained by solving the differential equation

$$L(h)w = q \qquad (4.65)$$

with some assigned boundary conditions. It corresponds to a variational principle according to which w minimizes the value of the functional

$$\int_\Omega \{\Pi(w, h) - 2qw\} \, d\Omega \to \min_w \qquad (4.66)$$

among admissible functions, that is, once continuously differentiable functions that may satisfy certain additional conditions.

Here Ω, h, q, L and Π denote in (4.65) and (4.66) the middle surface of a structural member, the distribution of thickness, the external loads, a positive-definite differential operator, and the strain energy density function, respectively. We shall consider the class of admissible loads $q(x) \geq 0$, and such that

$$\int_\Omega q(x)d\Omega \leq P_0. \qquad (4.67)$$

Here P_0 denotes a given constant value. As a measure of the rigidity of our thin-walled structure we shall take the quantity

$$J = J(h) = \max_q \max_{x \in \Omega} w(x, h, q). \qquad (4.68)$$

Let us prove a property of the criterion (4.68) with (4.67) that permits us to simplify computations of rigidity estimates.

Property 1. The load that causes a maximum deflection for any distribution of thickness (i.e., of the design) is a point load.

To prove this property, it suffices to consider two displacement functions w and w^c that correspond, respectively, to a distributed load in (4.67) and to a point load P_0, applied to the point x^1 for the deflection function w has its maximum value. The values of the functions w and w^c computed at the point x^1 are denoted w^1 and w^{c1}, respectively. Using (4.67), we obtain an estimate of the deflection function

$$\int_\Omega q w d\Omega \leq \left(\max_x w\right) \int_\Omega q d\Omega \leq w^1 P_0. \tag{4.69}$$

Now, using (4.69) and the minimum principle for the total energy, we arrive at a triple inequality

$$\int_\Omega \Pi(w^c, h) d\Omega - 2P_0 w^{c1} \leq \int_\Omega \Pi(w, h) d\Omega - 2P_0 w^1 \leq \int_\Omega \{\Pi(w, h) - 2qw\} d\Omega. \tag{4.70}$$

The energy relations

$$P_0 w^{c1} = \int_\Omega \Pi(w^c, h) d\Omega, \tag{4.71}$$

$$\int_\Omega q w d\Omega = \int_\Omega \Pi(w, h) d\Omega$$

expressing the equality between work performed by external loads and the strain energy stored permit us to simplify the inequality (4.70) to

$$P_0 w^{c1} \geq 2P_0 w^1 - \int_\Omega q w d\Omega \geq \int_\Omega q w d\Omega. \tag{4.72}$$

It follows from the inequality (4.72) that $w^{c1} \geq w^1$. Therefore, we have demonstrated that, for an arbitrary admissible load q satisfying (4.67), it is possible to indicate a point of application of a concentrated load P_0 that causes a maximal deflection not smaller than the maximal deflection caused by the distributed load $q(x)$. Thus, Property 1 has been proved.

Next, let us examine (4.65)–(4.68), if $J = J(h)$ stands for the maximal deflection of a plate. Let the maximum with respect to x in (4.68) occur at $x = x_1$ and the maximum with respect to q occur for $q = P_0 \delta(x - x_2)$ that is, for a point load applied at some point $x = x_2$.

Property 2. The maximum deflection $J = J(h)$ (where the maximum is taken with respect to both q and all $x \in \Omega$) is attained at the point of application of the point load, that is, for $x_1 = x_2$.

We shall prove this assertion by contradiction. Let us assume that, to the contrary, the maximum deflection is attained at some point x_1, while the concentrated load is applied at a different point x_2, where the deflection is equal to w_2. We also consider the deflection function w_a produced by applying the load P_0 at the point x. Let w_{a_1} denote the magnitude of deflection w_a evaluated at the point x_1. Using a well-known variational principle

$$\int_\Omega \Pi(w_a, h)d\Omega - 2P_0 w_{a_1} \leq \int_\Omega \Pi(w, h)d\Omega - 2P_0 w_1 \leq \int_\Omega \Pi(w, h)d\Omega - 2P_0 w_2,$$
(4.73)

using the energy relations

$$P_0 w_2 = \int_\Omega \Pi(w, h)d\Omega,$$
(4.74)

$$P_0 w_{a_1} = \int_\Omega \Pi(w_a, h)d\Omega,$$

we transform the inequalities (4.73) into the form

$$P_0 w_{a_1} \geq 2P_0 w_1 - \int_\Omega \Pi(w, h)d\Omega \geq \int_\Omega \Pi(w, h)d\Omega.$$
(4.75)

Looking at the right-hand side of inequality (4.75), we see that

$$P_0 w_1 - \int_\Omega \Pi(w, h)d\Omega \geq 0.$$
(4.76)

This estimate, combined with the left inequality in (4.75), implies that

$$w_{a_1} \geq w_1.$$

This contradiction completes the proof of Property 2.

Using these properties of criteria (4.68) with (4.67), we may restrict our discussion in estimating rigidity to applications of concentrated loads of magnitude P_0 and to computations of the deflections only at the points of application of these loads.

Chapter 5
Uncertainties in Fracture Mechanics and Optimal Design Formulations

Most investigations in the theory of optimal design of structures under strength constraints have been performed within a framework of the deterministic approach. That is, it is assumed that there is regular internal structure of material and that complete information is provided with regard to loading processes and boundary conditions. Corresponding optimal design formulations were typical for structures from elastic-plastic materials [Ban83, Pic88, HNT86, HA79, Ban81, Arm83, Aro89, EO83, HN88, HN96, Nei91, Hau81, HC81, Cea81, OR95, Pra72, Roz76, RK88, JM83, MU81].

Fewer studies have been devoted to an important class of problems involving brittle and quasi-brittle elastic body optimization, on the basis of modern fracture mechanics criteria [Ban97, Ban98, Ban99a, Ban99b, BBS05, BIM07, BIMS05a, BIMS05b, BMN00, BBB$^+$00, CF88, PH03, VHS02, TWK94, WK95, BPB$^+$00]. In accordance with fracture mechanics representations, it is necessary to consider all possibilities of crack appearance. In most cases, however, the number of cracks, their positions, orientations, sizes, and modes (opening cracks, shear cracks, ...) are unknown beforehand, so optimal design problems for brittle and quasi-brittle materials intrinsically contain uncertainty or randomness. Note that problems in the design of optimal structural systems with incomplete information are essentially different, both in the formulation and in research techniques. It is possible to consider various approaches to optimization problems with incomplete information such as the guaranteed approach, the probabilistic approach, or various mixed probabilistic-guaranteed approaches.

The guaranteed approach to optimal design of structures from brittle and quasi-brittle materials, considered in this book, is one of the feasible approaches to formulation and solution of these types of problems with incomplete information. In the guaranteed approach, it is assumed that a set Λ_ω containing all possible realizable vectors ω of unknown parameters (crack positions, orientations, sizes and others) is known, and that the shape of a structure having extremal cost functional and satisfying all strength constraints for all possible vectors of unknown parameters, contained in the realizable set

$$\omega \in \Lambda_\omega \tag{5.1}$$

N.V. Banichuk and P.J. Neittaanmäki, *Structural Optimization with Uncertainties*, Solid Mechanics and Its Applications 162, DOI 10.1007/978-90-481-2518-0_1,

is to be determined. Specifically, the objective is to find shapes of elastic bodies having smallest possible weight (volume or other cost functional), subject to appropriate constraints on stress intensity factors or energy release rates for static loads, and constraints on the number of cycles for fatigue cracks for cyclic loadings.

5.1 Basic Relations of Fracture Mechanics

Consider a deformed quasi-brittle elastic body occupying a two-dimensional region Ω with boundary Γ; the part Γ_v is taken as a design variable while the remaining part $\Gamma \setminus \Gamma_v$ is fixed. A surface traction T is given on the boundary Γ_σ, while zero displacement u is assigned on Γ_u, where $\Gamma = \Gamma_u + \Gamma_\sigma$. It is supposed that $\Gamma_v \subset \Gamma_\sigma$. The body contains a crack that is modeled by a rectilinear notch Γ_f. The boundaries of the notch are traction free. The state of a deformed body is described by a system of equations and boundary conditions of the theory of elasticity

$$\nabla \cdot \sigma = 0, \ \sigma = S : \varepsilon, \ \varepsilon = \tfrac{1}{2}\left(\nabla u + (\nabla u)^T\right), \tag{5.2}$$

$$(u)_{\Gamma_u} = 0, \ (\sigma \cdot n)_{\Gamma_\sigma} = T, \tag{5.3}$$

$$(\sigma \cdot n)_{\Gamma_f} = 0. \tag{5.4}$$

Here σ, ε, u, S and n denote, respectively, the stress tensor, the strain tensor, the displacement vector, the elastic modulus tensor (stiffness tensor), and the unit vector pointing in the direction of an outward normal to the boundary of the body. Basic relations of the theory of elasticity are derived under the assumptions that the laws of statics are obeyed, and deformations are small.

In perfectly brittle and quasi-brittle materials, it is expected that the conditions for a crack not to propagate are

$$g \equiv -\left\{\frac{\partial \Pi}{\partial l}\right\}_T < G_c, \tag{5.5}$$

$$\Pi = \frac{1}{2}\int_\Omega \varepsilon : S : \varepsilon d\Omega - \int_{\Gamma_\sigma} T \cdot u d\Gamma, \tag{5.6}$$

where Π is the total potential energy, l is the length of the crack, and subscript T denotes that the differentiation is performed under constant load. The criterion for fracture initiation for perfectly brittle and quasi-brittle materials can be written as

$$g = G_c, \tag{5.7}$$

where g is regarded as being applied, and, therefore, a function of geometry and load. On the other hand, G_c is regarded as a material parameter.

The energy release rate g and the stress intensity factors K_1 and K_2 occurring in asymptotic representations

$$\sigma_n = \frac{K_1}{\sqrt{2\pi r}} + O(1), \tag{5.8}$$

$$\sigma_{nt} = \frac{K_2}{\sqrt{2\pi r}} + O(1) \tag{5.9}$$

for stresses, determined near the crack tips, are connected by the formula [Che79, Hut79, KP85].

$$g = \frac{\beta}{E} \left(K_1^2 + K_2^2 \right). \tag{5.10}$$

Here, r denotes the distance of the considered point on the axis t of the local orthogonal coordinate system t, n from the crack tip. The axis t is parallel to the crack, while the origin is placed at the crack midpoint. Parameter $\beta = 1$ in plane stress state and $\beta = 1 - \nu^2$ in plane strain. Here ν is Poisson's ratio and E is Young's modulus. Note that brittle or quasi-brittle conditions for cracks not to propagate can be written in the form

$$K_i < K_{ic}, \ i = 1, 2 \tag{5.11}$$

for opening cracks ($i = 1$) and for shear cracks ($i = 2$), or in the form

$$\frac{\beta}{E} \left(K_1^2 + K_2^1 \right) < G_c \tag{5.12}$$

for the complex case, when the energy release rate and the stress intensity factors are characterized by opening and shear stresses. Here the values K_{ic} ($i = 1, 2$) and G_c are given brittle strength constants.

It is convenient to assess the strained state and determine the stress intensity coefficients if (following Bueckner) the representation for displacements, strains and stresses is written as

$$u = u^0 + u^f, \ \varepsilon = \varepsilon^0 + \varepsilon^f, \ \sigma = \sigma^0 + \sigma^f. \tag{5.13}$$

Here the quantities u^0, ε^0 and σ^0 define the stressed-strained state of a crack-free body, and are determined from the solution of the following boundary-value problem:

$$\nabla \cdot \sigma = 0, \ \sigma^0 = S : \varepsilon^0, \ \varepsilon^0 = \frac{1}{2} \left(\nabla u^0 + \left(\nabla u^0 \right)^T \right), \tag{5.14}$$

$$\left(u^0 \right)_{\Gamma_u} = 0, \ \left(\sigma^0 \cdot n \right)_{\Gamma_\sigma} = T. \tag{5.15}$$

The quantities u^f, ε^f and σ^f are the characteristics of the stressed-strained state of a cracked body that is not subjected to external loads on Γ_σ, with forces applied to

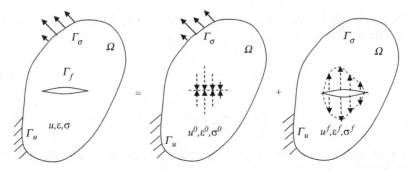

Fig. 5.1 Decomposition of actual displacements, strains and stresses into smooth and singular parts

the crack sides that are in opposite direction to forces arising in the corresponding areas of the crack-free body. The quantities u^f, ε^f, σ^f satisfy the boundary-value problem

$$\nabla \cdot \sigma^f = 0, \ \sigma^f = S : \varepsilon^f, \ \varepsilon^f = \frac{1}{2}\left(\nabla u^f + \left(\nabla u^f\right) T\right), \qquad (5.16)$$

$$\left(u^f\right)_{\Gamma_u} = 0, \ \left(n \cdot \sigma^f\right)_{\Gamma_\sigma} = 0, \qquad (5.17)$$

$$\left(n \cdot \sigma^f\right)_{\Gamma_f} = -\left(n \cdot \sigma^0\right)_{\Gamma_f}. \qquad (5.18)$$

Thus, the decomposition (5.13), shown in Fig. 5.1, separates the initial fields u, ε, σ into the smooth components u^0, ε^0 and σ^0, and the singular components u^f, ε^f and σ^f which define the stress intensity coefficients and the energy release rate during crack propagation.

It is important for the subsequent considerations that the possible locations of the cracks, arising from manufacture or exploitation, are unknown beforehand. This leads to essential complications caused by the necessity of analyzing a variety of crack locations and orientations and to solve corresponding boundary value problems (5.16)–(5.18). Significant simplifications are related to the possibility to solve the problem (5.16)–(5.18) effectively, and to obtain exact or approximate expressions for the stress intensity factors as a functions of loading and geometric parameters.

Let us assume that the length l of the rectilinear crack is small compared to the characteristic size L of the body ($l \ll L$) and that the crack is located far from the body boundaries. In this case the problem given by (5.16)–(5.18) admits of an exact (asymptotically exact) solution [KP85, Che79].

The stress intensity factors are given by

$$K_1 = \int_{-a}^{a} \sigma_n^f(t, 0) \left(\frac{2\pi(a + t)}{l(a - t)}\right)^{1/2} dt, \qquad (5.19)$$

$$K_2 = \int\limits_{-a}^{a} \sigma_{nt}^f(t,0) \left(\frac{2\pi(a+t)}{l(a-t)}\right)^{1/2} dt, \tag{5.20}$$

where $a = l/2$. If the stress fields in the problem for a crack-free body vary little along the crack length and can be regarded as approximately uniform:

$$\sigma_n^f(t,0) = \sigma_n^0 = \text{const}, \tag{5.21}$$

$$\sigma_{nt}^f(t,0) = \sigma_{nt}^0 = \text{const}, \tag{5.22}$$

then Eqs. (5.19)–(5.22) yield

$$K_1 = \sigma_n^0 \sqrt{\pi l/2}, \tag{5.23}$$

$$K_2 = \sigma_{nt}^0 \sqrt{\pi l/2}. \tag{5.24}$$

For a surface crack directed normally to the unloaded boundary of the body, the assumption that the crack length is much smaller that the radius of curvature R of the boundary, and the characteristic size L of the body ($l \ll R$ and $l \ll L$), i.e.

$$l \ll \min\{R, L\} \tag{5.25}$$

make it possible to reduce the analysis of the stressed-strained state of an elastic medium in the vicinity of the crack to the problem of the theory of elasticity of a half-plane with a cut at the surface. The boundary conditions at infinity correspond to the half-plane being exposed to tension in the s-direction parallel to the half-plane boundary. The coefficient K_1 is defined by [Hut79]

$$K_1 = \kappa \sigma_1^0 \sqrt{\pi l}, \quad \kappa = 1.12. \tag{5.26}$$

The expression (5.26) can be generalized to a surface crack in a strip of the width h, which is subjected to tensile stresses σ_s^0 applied at infinity, where the crack length may be comparable to the strip width h. In this case the quantity κ appearing in (5.26) is defined as [Hel84]

$$\kappa = \varphi(\lambda), \quad \lambda = l/h \ (h \sim l),$$

$$\varphi(\lambda) \equiv 1.12 - 0.23\lambda + 10.6\lambda^2 - 21.7\lambda^3 + 30.4\lambda^4. \tag{5.27}$$

In contrast to this, if the strip is subjected to stresses acting at infinity and corresponding to the case of pure-bending strain

$$\sigma_s = \sigma_s^0 \zeta/(h/2), \quad -h/2 \le \zeta \le h/2, \tag{5.28}$$

where ζ is the coordinate varying along the direction perpendicular to the strip boundaries, the relevant expression for K_1 is given by (5.26) with $\kappa = \psi(\lambda)$ [Hel84, Hut79], where

$$\psi(\lambda) \equiv 1.12 - 1.39\lambda + 7.3\lambda^2 - 13\lambda^3 + 14\lambda^4. \qquad (5.29)$$

The expressions (5.19)–(5.29) for the stress-intensity factors for surface and interior cracks can be effectively used in an optimization-based design.

5.2　Model Assumptions and Optimization Problems

Let us characterize the crack by the vector

$$\omega = \{x_c, y_c, l, \alpha\}, \qquad (5.30)$$

where (x_c, y_c) are coordinates of the crack mid-point, l and α are the length of the crack and the angle setting the crack orientation with respect to the global coordinate system x, y as shown in Fig. 5.2.

Consider the following model assumptions:

1. The perfect brittle or quasi-brittle body contains a crack, modeled by a rectilinear notch Γ_f.
2. The notch is traction free.
3. The length of considered cracks does not exceed the given limit value l_m.
4. The body has only one crack, but this crack can be characterized by any vector ω from given set Λ_ω.

Note that it is possible to admit the assumption 4 in the general case if the distances between the originating cracks are large enough to exclude the mutual influence of

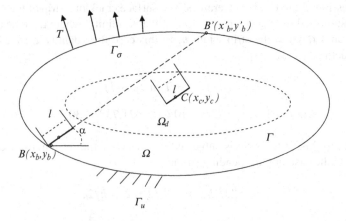

Fig. 5.2 Quasi-brittle body with cracks

cracks. In what follows some additional asymptotic assumptions concerning admissible variant of crack location are taken into account. For internal cracks, it is supposed that the limit length l_m of considered cracks does not exceed given parameter L_0

$$l \le l_m \ll L_0 = \min_{(x,y)} \min_{(x_c,y_c)} \left\{ (x - x_c)^2 + (y - y_c)^2 \right\}^{1/2} \tag{5.31}$$

where $(x, y) \in \Gamma$, $(x_c, y_c) \in \Omega_d$. This condition means that the cracks are located far from the boundaries of the body. The subdomain $\Omega_d \subset \Omega$ is considered as given.

For a surface crack, it is supposed that the crack is orthogonal to the boundary and the crack length is much smaller than the radius of curvature R of the boundary, and characteristic size L of the body

$$l \le l_m \ll \min\{R, L\},$$

$$L = \min_{(x_b,y_b)} L_{BB'}, \quad L_{BB'} = \left\{ (x_b - x_{b'})^2 + (y_b - y_{b'})^2 \right\}^{1/2}. \tag{5.32}$$

Here $B'(x_{b'}, y_{b'})$ is the point of intersection of the crack line and the boundary Γ (Fig. 5.2).

Accepted assumptions and available additional data concerning the most dangerous parts (subdomains) of the structure (in the sense of cracks appearance) allow us to consider the set Λ_ω of admissible cracks as given.

The problem of optimal structural design consists of finding the part of the boundary Γ_v, such that the cost functional J attains a minimum

$$J = J(\Gamma_v) \to \min_{\Gamma_v} \tag{5.33}$$

while satisfying strength constraints

$$\max_\omega g = \max_\omega \left\{ \frac{\beta}{E} \left(K_1^2 + K_2^2 \right) \right\} \le G, \tag{5.34}$$

where the maximum in (5.34) with respect to ω is found under the condition (5.1) and minimization in (5.33) is performed taking into account some additional geometric constraints written as

$$\Gamma_v \in H. \tag{5.35}$$

Here H is a given set of admissible curves.

Another possible formulation of optimization problems consists in finding an admissible shape of the contour Γ_v such that the maximum of the energy release rate g is minimized

$$g_* = \min_{\Gamma_v} \max_\omega \left\{ \frac{\beta}{E} \left(K_1^2 + K_2^2 \right) \right\} \tag{5.36}$$

Fig. 5.3 The body with a
surface crack

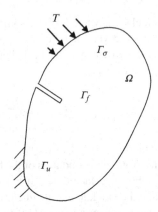

while satisfying the conditions (5.1), (5.35) and the isoperimetric constraint

$$\text{mes}\,\Omega = V, \tag{5.37}$$

where $v > 0$ is a given constant.

Up to now we have formulated the problems of optimal design with uncertainties concerning cracks for the static case when the loadings are fixed. There is a different scenario when we study optimal design of structural elements under cyclic loadings and fatigue crack growth. To formulate the problem of optimal design in this case, we reconsider a deformed elastic body (Fig. 5.3) occupying a two-dimensional domain Ω with boundary $\Gamma = \Gamma_u + \Gamma_\sigma$. Zero displacements are assigned on Γ_u, and surface tractions T are given on the boundary contour Γ_σ. The body contains a surface crack modeled by a rectilinear notch Γ_f. Making assumptions that the notch is traction free, we describe the state of a stressed and deformed body by the system of equations and boundary conditions of the theory of elasticity (5.2)–(5.4).

Suppose that the body is subjected to a cyclic loading, i.e.

$$T = T_0(x, y)p,$$
$$0 \leq p_{\min} \leq p \leq p_{\max}, \tag{5.38}$$

where $T_0(x, y)$ is a given amplitude loading function of the space coordinates, p is a given loading parameter, $p_{\min}$ and $p_{\max}$ are given positive values. For one cycle, the loading parameter increases from $p = p_{\min}$ up to $p = p_{\max}$ (loading process) and then decreases from $p = p_{\max}$ to $p = p_{\min}$ (unloading process). The loading and unloading processes are supposed to be quasistatic; dynamical effects are excluded. Let us apply the fatigue crack growth theory and suppose that the body contains an initial crack of length l_0. The process of fatigue crack growth under cyclic loading can be adequately characterized [Hel84, KP85] in the following form:

$$\frac{dl}{dn} = C(\Delta K)^m, \tag{5.39}$$

$$\Delta K = K_{\max} - K_{\min}. \tag{5.40}$$

Here C and m ($2 < m \leq 4$) are material constants, n is the number of cycles, K is a stress intensity factor, K_{max} and K_{min} are, respectively, the maximum and minimum values of the stress intensity factor K in any given loading cycle. The ordinary differential equation (5.39) defines the quasistatic process of crack growth, and determines the dependence of the crack length l on the number n. This equation is valid up to the moment when $l = l_{cr}$ and the unstable crack growth (fracture of the body) is attained. Suppose that the unstable crack growth is attained after $n = n_{cr}$ cycles, when the critical crack length l_{cr} satisfies the limiting relation

$$K(l_{cr}, \sigma_{max}) = K_{1c} \text{ or } l_{cr} = \varphi(\sigma_{max}, K_{1c}), \tag{5.41}$$

$$\sigma_{max} = p_{max}(\sigma_n^0)_{p=1}. \tag{5.42}$$

Here, φ is a given function and σ_{max} is the maximum stress in the uncracked body at the crack location.

Thus structural longevity can be measured by the number of load cycles

$$n = n_{cr} \tag{5.43}$$

for which $l = l_{cr}$ and the unstable fracture is realized. In the design process, the longevity constraint can be taken as

$$n_{cr} \geq n_0, \tag{5.44}$$

where N_0 is a given minimum value.

It is assumed for subsequent considerations that the possible location of initial cracks, arising from manufacture of exploitation, is unknown beforehand. In the context of structural design, this leads to essential complications in computation of n_{cr} cased by the necessity to analyze a variety of crack locations and orientations and to solve the corresponding structural analysis problems. To simplify the design process, the model assumptions 1, 2, 4, formulated above (in this section) can be used. The assumption 3 is replaced by the constraint that the initial length of the crack l_0 does not exceed a given limit value l_m that is much less than the radius of the surface curvature R and the characteristic size of the body L, i.e.

$$l_0 \leq l_m \leq l_{cr} \leq \min\{R, L\}. \tag{5.45}$$

This limit value l_m plays the role of a problem parameter.

Taking into account the incompleteness of information concerning the possible locations of initial cracks, we can rewrite the longevity constraint (5.44) in the following manner:

$$\min_{\omega} n_{cr} \geq n_0. \tag{5.46}$$

The optimization problem consists in finding the admissible part Γ_v of the boundary

$$\Gamma_v \subset \Gamma_\sigma \subset \Gamma \tag{5.47}$$

such that the quality functional (mass, volume, cost of the structure) is minimized

$$J_* = \min_{\Gamma_v} J(\Gamma_v) \tag{5.48}$$

while satisfying the longevity constraint (5.46) and (if necessary) some additional geometrical and mechanical constraints.

The shape of the body will be called optimal if for any body with a smaller quality functional, it is possible to select a vector of unknown parameters ω belonging to the admissible set Λ_ω such that some assigned constraints have been violated.

Another optimization problem consists in finding an admissible shape of the contour Γ_v such that the minimum critical number of cycles n_{cr} is maximized

$$n_* = \max_{\Gamma_v} \min_{\omega} n_{cr} \tag{5.49}$$

while satisfying the condition (5.1), and (if necessary) some additional geometrical or mechanical constraints

$$J(\Gamma_v) = J_0, \tag{5.50}$$

where $J(\Gamma_v)$ is a functional characteristics of the desired part of the boundary and J_0 is a given constant.

Chapter 6
Beams and Plates with Brittle-Fracture Constraints

6.1 Optimization of Beams

Consider the problem of optimal design of a beam taking into account the possibility of crack appearance at the beam surfaces. We assume that the beam of length L lies along the x-axis ($0 \leq x \leq L$) and that it has a rectangular cross-section with height $h = h(x)$ and constant width b.

Here we consider the statically determinate loading case (simply supported beams or cantilever beams subjected to the actions of transverse loads), with bending moment function $M(x)$ and shear load function $Q(x)$ acting on the cross-section of the beams. We regard these functions as known for $0 \leq x \leq l$. The maximum of stretching stresses is attained on the elongated beam surface for any cross-section $x = \xi$, i.e. $|\zeta| = h(\xi)/2$, where the coordinate ζ measures the distance from the center of the cross-section and varies in the interval $-h/2 \leq \zeta \leq h/2$. Therefore, we have

$$\max(\sigma_x) = 6\frac{|M(\xi)|}{bh^2(\xi)}. \qquad (6.1)$$

Suppose that

$$l_m \ll h_{\min} \leq \max h(x) \ll L, \qquad (6.2)$$

where $h_{\min}$ is a given positive constant. Use the expressions $K_1 = k\sigma_s^0\sqrt{\pi l}$, $k = 1.12$ [Hel84, Hut79] to evaluate the stress intensity factor K_1. The function $h = h(x)$ determining the shape of the beam is the unknown quantity. The problem of optimization consists in finding an admissible thickness distribution such that the volume of the beam is minimized, while satisfying strength and geometrical constraints

$$J = \int_0^L bh(x)dx \rightarrow \min_h, \qquad (6.3)$$

$$(K_1)_{\max} = \max_{\xi} \max_{l} \left(\frac{6|M(\xi)|\sqrt{\pi l(\xi)}}{bh^2(\xi)} \kappa \right) \leq K_{1C}, \qquad (6.4)$$

$$h(\xi) \geq h_{\min}, \quad l(\xi) \leq l_m. \qquad (6.5)$$

N.V. Banichuk and P.J. Neittaanmäki, *Structural Optimization with Uncertainties*, Solid Mechanics and Its Applications 162, DOI 10.1007/978-90-481-2518-0_1,
© Springer Science+Business Media B.V. 2010

The solution of the weight optimization problem for opening surface cracks is written in the explicit form

$$h(x) = \max \left\{ h_{\min}, \left(\frac{6\,|M(x)|\,\sqrt{\pi l_m}}{bK_{1C}} \kappa \right)^{1/2} \right\}. \tag{6.6}$$

The operation max in (6.6) leads to the selection of the larger of the two quantities inside the braces. If the maximum in (6.6) is realized for some subdomains by the second term inside the braces, we can show that the maximum of σ_x takes the constant value $K_{1C}/k\sqrt{\pi l_m}$ for these subdomains. On the remaining segments of the beam's length, where the height of the beam is constant: $h(x) = h_{\min}$, the max σ_x is smaller than $K_{1C}/k\sqrt{\pi l_m}$. Consequently, the applied minimax or "guaranteed" approach gives us the beam design with partially uniform strength if the admissible surface crack length is small enough.

Consider the optimal design of brittle and quasi-brittle beams with a low limiting value of shear intensity factor K_{2C}. Shear stresses are maximal at the neutral axis of the beam. We use the expression $\max(\sigma_{xy}) = 3\,|Q(\xi)|\,/2bh(\xi)$ to obtain a direct functional dependence of maximal shear when the crack is located at the x-axis and its length is much smaller than the minimum thickness of the beam, and smaller than the characteristic size of the domain of shear stress variation for the uncracked beam. In accordance with the formula 5.20 for K_2 we shall have

$$K_2 = 3\,|Q(\xi)|\,\sqrt{\pi l(\xi)}/2\sqrt{2}bh(\xi). \tag{6.7}$$

The problem of finding an optimal design consists in finding a thickness distribution from the condition of minimizing the beam volume (6.3) while satisfying the geometric conditions (6.5) and the fracture mechanics constraint

$$(K_2)_{\max} = \max_{\xi} \max_{l} \left(\frac{3\,|Q(\xi)|\,\sqrt{\pi l(\xi)}}{2\sqrt{2}bh\,(\xi)} \right) \leq K_{2C}, \tag{6.8}$$

where ξ is the coordinate of the crack midpoint. The solution of this problem can be written in the form

$$h(x) = \max \left\{ h_{\min}, \frac{3\,|Q(x)|\,\sqrt{\pi l_m}}{2\sqrt{2}bK_{2C}} \right\}. \tag{6.9}$$

Now consider the optimal design of a beam taking into account the possibility of crack appearance at the beam surfaces (mode I – opening cracks) and at the neutral line (mode II – shear cracks). Taking into account that the maximum of stretching stresses is attained on the elongated beam surface, and the shear stresses are maximum at the neutral axis of the beam, we suppose that the internal shear cracks are located on the x-axis, and the edge opening cracks arise at the beam surfaces.

The problem of finding an optimal thickness distribution consists in volume minimization, and has the following solution:

$$h(x) = \max \left\{ k \frac{6|M(x)| \sqrt{\pi l_m}}{b K_{1C}}, \frac{3|Q(x)| \sqrt{\pi l_m}}{2\sqrt{2}b K_{2C}}, h_{min} \right\}. \qquad (6.10)$$

6.2 Optimum Shapes of Holes in Elastic Plates

Let us consider the problem arising when we wish to determine optimum shapes of holes in infinite plates with surface cracks having the smallest stress intensity factor. Assume that the admissible length of the cracks satisfies the constraint $l \leq l_m$, the boundary of the hole Γ_ν has no external loads applied to it, and the plate is subjected to tensile forces applied at infinity, i.e.

$$(\sigma_n)_{\Gamma_\nu} = (\sigma_{ns})_{\Gamma_\nu} = 0,$$
$$(\sigma_x)_\infty = \sigma_1^\infty, (\sigma_y)_\infty = \sigma_2^\infty, \qquad (6.11)$$

where σ_1^∞ and σ_2^∞ are given positive constants and n and s denote, respectively, the directions normal and tangent to the curve Γ_ν. We suppose that the hole contour is smooth, cracks are orthogonal to the contour, and their lengths are much smaller than the hole curvature radius, i.e. $l \leq l_m \ll \min R(P)$.

Here the minimum is taken with respect to $P \in \Gamma_\nu$. The optimization problem consists in finding the admissible shape of the contour Γ_ν such that the maximum of the stress intensity factor K_1 is minimized

$$J_* = \min_{\Gamma_\nu} \max_{P \in \Gamma_\nu} \max_{l \leq l_m} K_1. \qquad (6.12)$$

Conventional decomposition reduces the analysis of stressed and deformed states to the solution of the following two boundary value problems. First we solve a boundary value problem of the plane theory of elasticity for an uncracked plate with a hole and loading conditions at the infinity. The solution of this problem is approximated by homogeneous stresses fields in the vicinity of the boundary point P (tension stresses in tangent to the curve Γ_ν direction). Then we solve a second problem for the plate with a hole and crack, in which boundaries are loaded by tractions arising in the first problem $\sigma_s = -\sigma_s^0$, and obtain the expression for the stress intensity factor $K_1 = k\sigma_s^0 \sqrt{\pi l}$ ($k = 1.12$). The problem (6.12) takes the form

$$J_* = k\sqrt{\pi l_m} \min_{\Gamma_\nu} \max_{P \in \Gamma_\nu} \sigma_s^0(P). \qquad (6.13)$$

Thus the optimization problem for a plate with a hole and surface cracks is reduced to the problem of tension stresses minimization (or stress intensity minimization); the solution of this problem is presented in [Ban83, Ban90].

6.3 Optimal Design of Bimaterial Layered Beams

Suppose the beam lies along the x-axis and that it has length L and rectangular cross-section with constant (along the x-axis) height h and width b. The beam is supposed to consist of layers, with heights h_1, h_2:

$$h = h_1 + h_2,$$

cross-section areas S_1, S_2:

$$S = S_1 + S_2$$

Young's moduli E_1, E_2 and specific weights γ_1, γ_2 (per unit volume). The neutral line of the beam coincides with the x-axis. The beam is subjected to pure bending in the xy-plane. The cross-section composed of two rectangular areas S_1, S_2 is rotated with respect to the z-axis perpendicular to the plane of bending and passing through the neutral axis of the beam. The distances between the neutral line ($y = 0$) and the outer surfaces ($y = $ const in plane of bending xy) of the beam are denoted by H_1 and H_2:

$$h = H_1 + H_2.$$

The values H_1 and H_2 are determined with the help of the condition of zero resultant force applied to the beam cross-section

$$0 = \int_S \sigma_x dS = \frac{E_1}{R} \int_{S_1} y dS + \frac{E_2}{R} \int_{S_2} y dS, \qquad (6.14)$$

where σ_x, R are respectively the stress component and the curvature radius.

Using this relation and performing necessary transformations, we will have

$$H_1 = \frac{E_1 h_1^2 + E_2 h_2 (h_2 + 2h_1)}{2(E_1 h_1 + E_2 h_2)}, \qquad (6.15)$$

$$H_2 = \frac{E_2 h_2^2 + E_1 h_1 (h_1 + 2h_2)}{2(E_1 h_1 + E_2 h_2)}. \qquad (6.16)$$

The equilibrium condition, written for the bending moment, gives us to the following relation:

$$M = \int_S y \sigma_x dS = E_1 \frac{I_1}{R} + E_{12} \frac{I_2}{R}. \qquad (6.17)$$

Here I_1 and I_2 are the moments of the inertia of cross-sectional areas S_1, S_2 with respect to the z-axis. Hence the stress distribution is described by the formulas

$$\sigma_x = E_1 \frac{y}{R} = \frac{E_1 M y}{E_1 I_1 + E_2 I_2}, \quad -H_1 \le y \le h_1 - H_1,$$

$$\sigma_x = E_2 \frac{y}{R} = \frac{E_2 M y}{E_1 I_1 + E_2 I_2}, \quad h_1 - H_1 \le y \le H_2. \tag{6.18}$$

The maximum of the absolute value of the considered stresses is attained either at

$$y = -H_1, \ \sigma_1 = |\sigma_x|$$

or else at

$$y = H_2, \ \sigma_2 = |\sigma_x|.$$

We have the following expression to compute the values σ_1 and σ_2:

$$\sigma_1 = \frac{E_1 H_1 M}{E_1 I_1 + E_2 I_2}, \ \sigma_2 = \frac{E_2 H_2 M}{E_1 I_1 + E_2 I_2}, \tag{6.19}$$

$$I_1 = \frac{b}{3} \left[H_1^3 + (h_1 - H_1)^3 \right], \tag{6.20}$$

$$I_2 = \frac{b}{3} \left[H_2^3 + (h_2 - H_2)^3 \right], \tag{6.21}$$

$$E_1 I_1 + E_2 I_2 = \frac{b}{3} \left(E_1 h_1^3 + E_2 h_2^3 \right) + b E_1 h_1 H_1 (H_1 - h_1) + b E_2 h_2 H_2 (H_2 - h_2). \tag{6.22}$$

We allow the possibility of normal-tear cracks starting at the surfaces $y = -H_1$ and $y = H_2$ of the beam. It is assumed also that the lengths of the rectilinear cracks do not exceed a given limiting value l_m, i.e.

$$l \le l_m \ll \min(h_1, h_2). \tag{6.23}$$

Furthermore, it is presumed that the crack direction is perpendicular to the boundary. In this case the expression for evaluation of the maximum value of the stress intensity factor can be written in explicit form as

$$K_1 = \kappa \sigma_i \sqrt{\pi l_m}, \ (\kappa = 1.12, \ i = 1, 2). \tag{6.24}$$

The problem of optimal beam design consists in finding admissible heights h_1, h_2 that provide the minimum of the beam-weight function

$$J = (\gamma_1 h_1 + \gamma_2 h_2) b L \tag{6.25}$$

and satisfies both the strength constraints

$$(K_1)_i \le (K_{1C})_i, \ i = 1, 2, \tag{6.26}$$

and the geometrical conditions

$$h_1 \geq h_1^0, \; h_2 \geq h_2^0, \tag{6.27}$$

where $h_1^0 > 0$, $h_2^0 > 0$ are given parameters and the given brittle-strength constants are designated as $(K_{1C})_1$, $(K_{2C})_2$.

For the sake of convenience, we shall introduce the following dimensionless variables and notations:

$$\widetilde{h_1} = \frac{h_1}{h_1^0}, \; \widetilde{h_2} = \frac{h_2}{h_1^0}, \; t = \frac{\widetilde{h_2}}{\widetilde{h_1}},$$

$$k = \frac{h_2^0}{h_1}, \; \alpha = \frac{E_2}{E_1}, \; r = \frac{\gamma_2}{\gamma_1}, \tag{6.28}$$

$$\widetilde{J} = \frac{J}{bLh_1^0\gamma_1}, \; \beta_i = \left[\frac{\kappa M \sqrt{\pi l_m}}{b(h_1^0)^2 (K_{1C})_i} \right]^{1/2}.$$

With these dimensionless variables (with tilde omitted) the expression for the cost function of the optimization, (problem (6.25)–(6.27) and constraints can be written as

$$J = h_1(1 + rt), \tag{6.29}$$

$$h_1 \geq 1, \; h_1 \geq k\psi(t), \tag{6.30}$$

$$h_1 \geq \beta_1 \psi_1(\alpha, t), \; h_2 \geq \beta_2 \psi_2(\alpha, t). \tag{6.31}$$

Here

$$\psi(t) = t^{-1}, \; \psi_1(\alpha, t) = \varphi_1(\alpha, t)\varphi(\alpha, t),$$

$$\psi_2(\alpha, t) = \varphi_2(\alpha, t)\varphi(\alpha, t),$$

$$\varphi_1(\alpha, t) = \left[6(1 + 2\alpha t + \alpha t^2) \right]^{1/2}, \tag{6.32}$$

$$\varphi_2(\alpha, t) = \left[6\alpha(1 + 2\alpha t + \alpha t^2) \right]^{1/2},$$

$$\varphi(\alpha, t) = \left(1 + 4\alpha t + 6\alpha t^2 + 4\alpha t^3 + \alpha^2 t^4 \right)^{-1/2}.$$

Using the functions in (6.32) we can formulate the optimization problem in compact form

$$J = h_1(1 + rt) \to \min_{h_1, t}, \tag{6.33}$$

$$h_1 \geq \Psi, \tag{6.34}$$

where

$$\Psi \equiv \max \left\{ 1, k\psi(t), \beta_1 \psi_1(\alpha, t), \beta_2 \psi_2(\alpha, t) \right\}. \tag{6.35}$$

The operation max in (6.35) leads to the selection of the larger of the four quantities inside the braces. If the maximum in (6.35) is realized for the first or the second term we can conclude that the corresponding geometric constraint from (6.27) is active for optimal design.

If the maximum in (6.35) is realized for the third of the fourth term it means that the fracture mechanics constraint (6.26) is active for $y = -H_1$ or for $y = H_2$. Functions

$$h_1 = 1, \ h_1 = k\psi(t), \ h_1 = \beta_1\psi_1(t), \ h_1 = \beta_2\psi_2(\alpha, t)$$

and dependences

$$h = J/(1 + rt), \ J = 2, 3$$

are shown in Fig. 6.1 by the lines 1, 2, ..., 6 for the case

$$r = 1, \ \alpha = 0.5, \ k = 1, \ \beta_1 = 1, \ \beta_2 = 2$$

and in Fig. 6.2 for the case $r = 0.5$ (other parameters are the same as for the previous case). If we exclude the geometric constraints from consideration we reduce the previous optimization problem (6.33)–(6.35) to the new one with

$$\Psi \equiv \max\{\beta_1\psi_1(\alpha, t), \beta_2\psi_2(\alpha, t)\}. \tag{6.36}$$

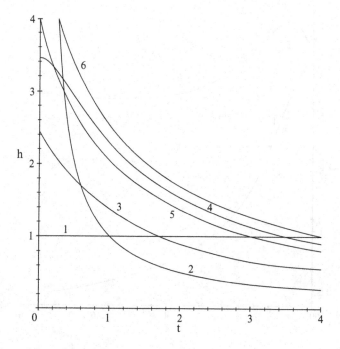

Fig. 6.1 Dependences of h on t for $r = 1$

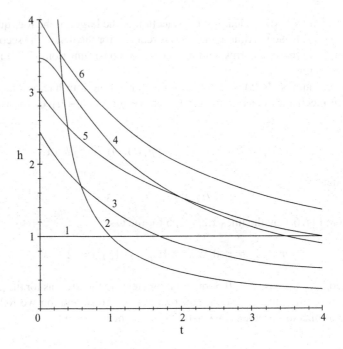

Fig. 6.2 Dependences of h on t for $r = 0.5$

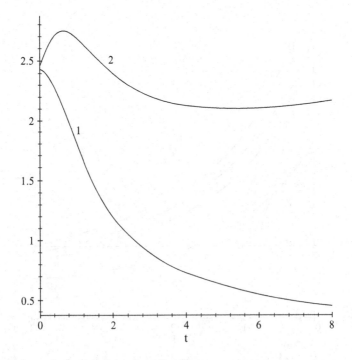

Fig. 6.3 Active strength constraint and cost function

The optimal solution

$$t_* = 5.22087, \ J_* = 2.109988$$

of the problem (6.33)–(6.35) was obtained for the case $r = 0.5, \alpha = 0.1, \beta_1 = 1,$ $\beta_2 = 2$. The active strength constraint and the values of the cost function J at the point of the constraint curve are shown in Fig. 6.3 by the curves 1, 2 for $0 \le t \le 8$.

Chapter 7
Optimization of Axisymmetric Shells Against Brittle Fracture

7.1 Basic Relations of the Membrane Shell Model

Consider a shell which has the shape of a surface of revolution, the axis of which coincides with the x-axis (Fig. 7.1).

The position of the meridian plane is specified by the angle θ, which is measured from a certain fixed meridian plane; the position of the parallel circle is defined by the angle φ between the normal to the surface and the axis of rotation, $r = r(x)$ is the radius of the parallel circle, which determines the distance from a point on the neutral surface of the shell to its axis of rotation and $0 \leq x \leq L$, where L is the specified length of the shell. The quantities $r(0) = r_1$ and $r(L) = r_2$, which correspond to the ends of the shell, are assumed to be given and to satisfy the inequalities $r_1 \geq 0, r_2 \geq 0$. The meridian plane and the plane which is perpendicular to the meridian are the planes of principal curvatures of the surface of the shell at the point being considered. The corresponding principal radii of curvature are r_φ and r_θ. We will use the following relations between meridional curvature radius r_φ, circumferential curvature radius r_θ and radius r

$$r_\varphi = -\frac{\left(1 + \left(\frac{dr}{dx}\right)^2\right)^{3/2}}{\frac{d^2r}{dx^2}}, \tag{7.1}$$

$$r_\theta = r\left(1 + \left(\frac{dr}{dx}\right)^2\right)^{1/2}. \tag{7.2}$$

The thickness and the radius distributions

$$h = h(\varphi) = h(\varphi(x)), \; r = r(\varphi) = r(\varphi(x)) \tag{7.3}$$

are assumed to satisfy the well-known conditions of the theory of thin elastic shells [TW59, Flu73]

$$h \leq h_m = \max_\varphi h(\varphi) = \max_x h(\varphi(x)) \ll r_m, \tag{7.4}$$

N.V. Banichuk and P.J. Neittaanmäki, *Structural Optimization with Uncertainties*, Solid Mechanics and Its Applications 162, DOI 10.1007/978-90-481-2518-0_1, © Springer Science+Business Media B.V. 2010

Fig. 7.1 Element of the
axisymmetric shell

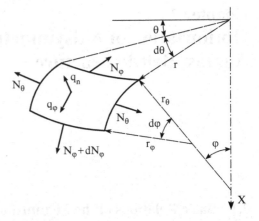

Fig. 7.2 Portion of the shell
above parallel circle *AB*

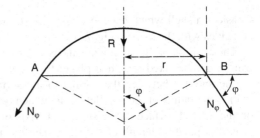

where

$$r_m = \min \left\{ \min_{\varphi} r_\varphi(\varphi), \min_{\varphi} r_\theta(\varphi) \right\} = \min \left\{ \min_{x} r_\varphi(\varphi(x)), \min_{x} r_\theta(\varphi(x)) \right\}. \quad (7.5)$$

The maxima and minima with respect to x in (7.4), (7.5) are defined at the interval
$[0, L]$ and the external operation min in front of the braces denotes the search for
the smaller of the two quantities.

The shell is loaded by axisymmetric forces in the meridian planes. The intensities of the external loads, which act in the directions normal and tangential to the
meridian, are denoted by q_n and q_φ (see Fig. 7.1). The equilibrium equations can be
written in the form [Tim56, TW59]

$$\frac{N_\varphi}{r_\varphi} + \frac{N_\theta}{r_\theta} = q_n, \quad (7.6)$$

$$2\pi r N_\varphi \sin \varphi + R = 0. \quad (7.7)$$

Here N_φ, N_θ are the magnitudes of the normal membrane forces per unit length
which can be determined from the equilibrium equations, and R is the resultant of
the total load on the portion of the shell above the parallel circle (Fig. 7.2), defined
by the angle φ.

For the normal stresses arising in the shell we have

$$\sigma_\varphi = \frac{N_\varphi(\varphi)}{h(\varphi)}, \quad \sigma_\theta = \frac{N_\theta(\varphi)}{h(\varphi)}, \tag{7.8}$$

where the thickness of the shell $h = h(\varphi)$ can be varied in the meridian direction. The shear stress $\sigma_{\varphi\theta} = 0$ because the shape of the shell, thickness distribution and loads are axisymmetric.

It is seen from (7.6), (7.7) that for given functions

$$q_n(\varphi), R(\varphi), r_\varphi(\varphi), r_\theta(\varphi), r(\varphi)$$

the force N_φ is obtained from Eq. (7.7), and the force N_θ is determined with the help of Eq. (7.6). We have

$$N_\varphi = -\frac{R}{2\pi r \sin \varphi}, \tag{7.9}$$

$$N_\theta = r_\theta \left(\frac{R}{2\pi r r_\varphi \sin \varphi} + q_n \right). \tag{7.10}$$

It is assumed that a through crack can arise in the shell during its manufacture or exploitation. It is also supposed that the material of the shell is quasi-brittle and the crack is rectilinear. The crack length l is larger than h_m, and is small with respect to the characteristic size r_m of the shell, i.e.

$$h_m \ll l \ll r_m. \tag{7.11}$$

The size, location and orientation of the crack are not fixed beforehand.

The condition for the crack not to propagate, as is well known from quasi-brittle fracture mechanics [KP85, Par92, SL68, Smi91], is $K_1 < K_{1C}$. Here K_1 is the stress intensity factor in the opening mode, and K_{1C} is a given quasi-brittle strength constant (toughness of material). This condition can be approximated by the inequality

$$K_1 \leq K_{1\varepsilon} \tag{7.12}$$

if we introduce a small positive parameter ε with the meaning [Bol61, Str47]

$$K_{1\varepsilon} = K_{1C} - \varepsilon(\varepsilon > 0) \tag{7.13}$$

of limiting value of stress intensity factor. This is made for convenience in the optimization procedure.

Note that it is possible to use the following expression [Che79, Hut79, Hel84] for the stress intensity factor of the through crack:

$$K_1 = \begin{cases} \sigma_n^0 \sqrt{\pi l/2} & \text{if } \sigma_n^0 \geq 0 \\ 0 & \text{if } \sigma_n^0 \leq 0 \end{cases} \tag{7.14}$$

when the through crack is small enough and is distant from the shell boundaries. Here σ_n^0 is the normal stress in the uncracked shell at the crack location. Subscript n means that the stress acts in the direction normal to the crack banks. It is assumed here that

$$\max \sigma_n^0 > 0, \tag{7.15}$$

where max operation in (7.15) is performed for all possible positions and orientations of the crack. If this assumption is violated, that is global compression takes place, then the opening cracks cannot be realized ($K_1 = 0$). In this case a fracture mechanics constraint is inessential.

To describe all considered variants of the crack appearance in the shell we shall characterize the crack by the vector

$$\omega = \{\varphi_C, l, \alpha\}, \tag{7.16}$$

where φ_C is the coordinate of the crack midpoint, l and α are the length of the crack and the angle setting the crack orientation with respect to the meridian. If $\alpha = 0$ the crack is oriented in meridian direction (axial crack) and for $\alpha = \pi/2$ the crack is oriented in the parallel direction. Note that the second coordinate θ_C of the crack midpoint is inessential and is omitted because we consider the axisymmetric problems in the frame of a guaranteed approach and admit all locations of the crack in parallel direction ($0 \leq \theta_C \leq 2\pi$). It is supposed that the length l of the crack does not exceed the given limit value l_m, where $l \leq l_m \ll r_m$.

Taking into account accepted assumptions and available additional data concerning the most dangerous parts (subdomains) of the structures (in the sense of cracks appearance), it is possible to consider the set Λ_ω ($\omega \in \Lambda_\omega$) as given.

We will consider the problem of the material volume minimization, taking the shell thickness $h = h(\varphi)$ and the shell radius $r = r(\varphi)$ as design variables. We will study variants of the problem in which the thickness distribution $h(\varphi)$ or the radius distribution $r(\varphi)$ or both are found in the context of material volume or weight minimization. Thus it is required to find the optimal thickness distribution $h = h(\varphi)$ and/or the radius distribution $r = r(\varphi)$ that satisfies the inequality (7.12), for any admissible crack location, orientation and size l ($l \leq l_m$, l_m is the given limit value of the considered cracks) and minimize the functional

$$J = \int_0^{2\pi} \int_{\varphi_0}^{\varphi_f} h r_\varphi r_\theta \sin \varphi \, d\varphi \, d\theta = 2\pi \int_{\varphi_0}^{\varphi_f} h r_\varphi r_\theta \sin \varphi \, d\varphi \to \min \tag{7.17}$$

under the geometric constraint

$$\chi(h, r) \geq 0, \tag{7.18}$$

where χ is a given function or functional. This means that we will use the guaranteed (minimax) approach for formulation and solution of the structural optimization problems, taking into account all admissible possibilities of the crack appearance.

The reformulated optimization problem consists in finding the thickness distribution $h(\varphi)$ and/or the radius $r(\varphi)$, such that the cost functional (7.17) attains a minimum while satisfying the geometrical constraint (7.18) and the strength constraint

$$\max_{\omega} K_1 \leq K_{1\varepsilon}, \qquad (7.19)$$

where the maximum with respect to ω is found under the condition

$$\omega \in \Lambda_\omega \equiv \left\{ \varphi_0 \leq \varphi \leq \varphi_f, \ 0 < l \leq l_m, \ 0 \leq \alpha \leq \frac{\pi}{2} \right\}. \qquad (7.20)$$

Using the explicit representation (7.14) for K_1, and remembering that σ_n^0 does not depend on the length of the crack, we conclude that the maximum of K_1 with respect to l is realized when $l = l_m$.

The following property holds: the maximum of K_1 with respect to α is attained when α takes one of two values: $\alpha = 0$ (axial crack) or $\alpha = \pi/2$ (peripheral crack).

To prove this statement, we note that in the membrane theory of thin shells of revolution we have three non-zero components of the stress tensor: two normal stresses σ_φ, σ_θ and one shear stress $\sigma_{\varphi\theta}$. Taking into account that the shell shape and the applied external loads are axially symmetrical, and using equilibrium equations, we see that the shear stress $\sigma_{\varphi\theta}$ is zero and the only non-zero normal stresses are σ_φ, σ_θ [Tim56, TW59]. This means that the membrane normal stresses σ_φ, σ_θ are principal and, consequently, we have

$$\begin{cases} \sigma_\varphi \leq \sigma_n^0 \leq \sigma_\theta & \text{if} \quad \sigma\varphi \leq \sigma_\theta, \\ \sigma_\theta \leq \sigma_n^0 \leq \sigma_\varphi & \text{if} \quad \sigma\theta \leq \sigma_\varphi. \end{cases} \qquad (7.21)$$

Thus we have proved the formulated inequalities. Using the minimax approach (construction of optimal solution for the worst case), we need to consider only axial cracks and peripheral cracks.

If the inequality (7.15) is satisfied, we can use the following expression for the left hand side of the inequality 7.19:

$$\max_{\omega} K_1 = \begin{cases} \sigma_\varphi \sqrt{\pi l_m/2}, & 0 < \sigma_\varphi \geq \sigma_\theta, \\ \sigma_\theta \sqrt{\pi l_m/2}, & 0 < \sigma_\theta \geq \sigma_\varphi, \\ 0, & \sigma_\varphi \leq 0 \text{ and } \sigma_\varphi \leq 0, \end{cases} \qquad (7.22)$$

where

$$\sigma_\theta = \left(\sigma_n^0 \right)_{\alpha=0}, \ \sigma_\varphi = \left(\sigma_n^0 \right)_{\alpha=\pi/2}. \qquad (7.23)$$

Consequently, the inequality (7.19) can be written as a system of two inequalities

$$\max_{\varphi} \left(\frac{\sqrt{\pi l_m/2}}{h} N_\varphi \right) \le K_{1\varepsilon}, \tag{7.24}$$

$$\max_{\varphi} \left(\frac{\sqrt{\pi l_m/2}}{h} N_\theta \right) \le K_{1\varepsilon}. \tag{7.25}$$

Here N_φ and N_θ are evaluated with the help of the expressions (7.9), (7.10).

7.2 Some Problems of Optimal Thickness Distribution

Suppose that the middle surface of the shell is prescribed, i.e. $r = r(\varphi)$ and, consequently, $r_\varphi = r_\varphi(\varphi)$, $r_\theta = r_\theta(\varphi)$ are given (see (7.1), (7.2)). The thickness distribution is considered as an unknown desired design variable $h = h(\varphi)$.

To satisfy (7.24), (7.25), it is necessary and sufficient to require that $h(\varphi) \ge \frac{\sqrt{\pi l_m/2}}{K_{1\varepsilon}} \max \left\{ -\frac{R(\varphi)}{2\pi r(\varphi)\sin\varphi}, r_\theta(\varphi) \left(\frac{R(\varphi)}{2\pi r(\varphi) r_\varphi(\varphi)\sin\varphi} + q_n(\varphi) \right) \right\}$ for any $\varphi \in [\varphi_0, \varphi_f]$. The operation max selects the larger of the two quantities inside the braces. Now the original problem (7.17), (7.19), (7.20) with the geometric constraint (7.18) written as

$$\chi \equiv h(\varphi) - h_0 \ge 0 \tag{7.26}$$

is reduced to a problem with explicit inequalities, superposed on the desired function $h(\varphi)$; it has the following analytical solution:

$$h = \max \left\{ h_0, \frac{\sqrt{\pi l_m/2}}{K_{1\varepsilon}} N_\varphi, \frac{\sqrt{\pi l_m/2}}{K_{1\varepsilon}} N_\theta \right\} =$$

$$= \max \left\{ h_0, -\frac{R}{2\pi r \sin\varphi} \frac{\sqrt{\pi l_m/2}}{K_{1\varepsilon}}, r_\theta \frac{\sqrt{\pi l_m/2}}{K_{1\varepsilon}} \left(\frac{R}{2\pi r_\varphi \sin\varphi} + q_n \right) \right\}. \tag{7.27}$$

For any $\varphi \in [\varphi_0, \varphi_f]$ the max operation in (7.26) means finding the maximum of the three values in braces.

In what follows we consider particular problems of optimal shell design and illustrate the solutions. Before starting with examples we make one note. For unclosed (open) shells supported usually by some rings or other arrangements against circumferential extension, some bending will occur near the supports. However, the edge effect is localized and the edge zone with meaningful bending moments is relatively small. At a certain distance from the boundary we can use the membrane theory with satisfactory accuracy [TW59].

In this section we consider particular examples of optimal design of toroidal, conical and spherical shells. Detailed analysis of stresses arising in these shells can be found in [Flu73, TW59].

Fig. 7.3 Toroidal tank

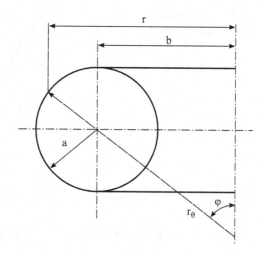

7.2.1 Thickness Distribution for Toroidal Shell

Consider a shell in the form of a torus obtained by rotating a circle of radius a about a vertical axis (Fig. 7.3 shows a half of the shell). The shell is subject to uniform pressure p. The forces N_φ, N_θ are obtained by considering the equilibrium of the ring – shaped element of the shell and written as

$$N_\varphi = \frac{pa(r+b)}{2r}, \quad N_\theta = \frac{pa}{2}. \tag{7.28}$$

Taking into account that $N_\varphi > N_\theta$, we will have optimal thickness distribution in the following form:

$$h = \max\left\{h_0, \frac{\sqrt{\pi l_m/2}\, pa(r+b)}{2rK_{1\varepsilon}}\right\}. \tag{7.29}$$

It is seen from (7.29) that the thickness decreases when the radius r increases.

7.2.2 Thickness Distribution for Conical Shell

Consider a conical tank filled with a liquid of specific weight γ (Fig. 7.4). Denote the distance from the bottom and the total depth of the liquid by y and d respectively.

If we substitute the expressions for the pressure q_n, the load R (weight of the liquid above the parallel circle AB defined by the angle φ) and corresponding geometrical values of the conical shell

Fig. 7.4 Conical tank filled
with a liquid

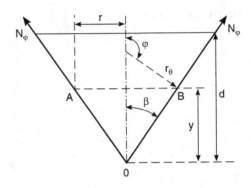

$$q_n = \gamma(d - y), \ R = -\pi\gamma y^2 \left(d - \frac{2}{3}y\right) \tan^2 \beta,$$

$$\varphi = \frac{\pi}{2} + \beta, \ r_\varphi = \infty, \ r = y \tan \beta \tag{7.30}$$

into the expressions (7.9), (7.10), we will have

$$N_\theta = \frac{q_n r}{\sin \varphi} = \frac{\gamma(d - y)y \tan \beta}{\cos \beta},$$

$$N_\varphi = -\frac{R}{2\pi r \cos \beta} = \frac{\gamma y \left(d - \frac{2}{3}y\right) \tan \beta}{2 \cos \beta}. \tag{7.31}$$

Equation (7.31) implies

$$N_\theta \geq 0, \ N_\varphi \geq 0 \tag{7.32}$$

for any $y \in [0, d]$, and

$$N_\theta - N_\varphi = \frac{\gamma y \left(d - \frac{4y}{3}\right) \tan \beta}{2 \cos \beta} \tag{7.33}$$

is positive for $0 \leq y \leq 3d/4$ and negative for $3d/4 < y \leq d$. We have

$$N_\theta \geq N_\varphi, \ 0 \leq y \leq \frac{3}{4}d,$$

$$N_\theta \leq N_\varphi, \ \frac{3}{4}d < y \leq d. \tag{7.34}$$

The optimal design of the conical shell can be represented as

$$h = \max\left\{h_0, \frac{\gamma \tan\beta \sqrt{\pi l_m/2}}{\cos\beta K_{1\varepsilon}}(d-y)y\right\}, \quad 0 \le y \le \frac{3}{4}d,$$

$$h = \max\left\{h_0, \frac{\gamma \tan\beta \sqrt{\pi l_m/2}}{2\cos\beta K_{1\varepsilon}}\left(d-\frac{2}{3}y\right)y\right\}, \quad \frac{3}{4}d < y \le d. \tag{7.35}$$

Note that the maximum value of the force N_θ

$$(N_\theta)_{\max} = (N_\theta)_{y=d/2} = \frac{\gamma d^2 \tan\beta}{4\cos\beta}$$

is larger than the maximum value of the force N_φ

$$(N_\varphi)_{\max} = (N_\varphi)_{y=3d/4} = \frac{3\gamma d^2 \tan\beta}{16\cos\beta}.$$

Consequently, if the parameter h_0 is sufficiently small, the thickness distribution $h(y)$ attains its maximum at $y = d/2$.

7.2.3 Thickness Distribution for Spherical Shell

Consider a spherical tank filled with liquid of a specific weight γ and supported along a parallel circle AB (Fig. 7.5).

The inner pressure q_n, the resultant R of this pressure for the portion of the shell, and the geometrical parameters can be written as

$$r_\varphi = r_\theta = a, \quad r = a\sin\varphi, \tag{7.36}$$

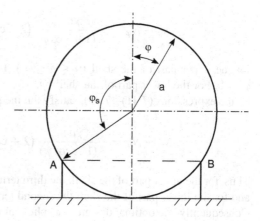

Fig. 7.5 Spherical tank filled with a liquid

$$q_n = \gamma a (1 - \cos \varphi),$$ (7.37)

$$R = -2\pi a^2 \int_0^\varphi \gamma a (1 - \cos \varphi) \sin \varphi \cos \varphi d\varphi =$$ (7.38)

$$= -2\pi a^3 \gamma \left[\frac{1}{6} - \frac{1}{2} \cos^2 \varphi \left(1 - \frac{2}{3} \cos \varphi \right) \right],$$

where a is the radius of the sphere.

Corresponding expressions for N_φ and N_θ are given by the following formulas:

$$N_\varphi = -\frac{R}{2\pi r \sin \varphi} = \frac{\gamma a^2}{6} \left(1 - \frac{2 \cos^2 \varphi}{1 + \cos \varphi} \right),$$ (7.39)

$$N_\theta = r_\theta \left(\frac{R}{2\pi r r_\varphi \sin \varphi} + q_n \right) = \frac{\gamma a^2}{6} \left(5 - 6 \cos \varphi + \frac{2 \cos^2 \varphi}{1 + \cos \varphi} \right).$$ (7.40)

The expressions (7.38)–(7.40) for R, N_φ and N_θ are valid for the upper portion of the tank ($\varphi < \varphi_s$). For the lower portion of the tank ($\varphi > \varphi_s$) we have the following expressions for R, N_φ and N_θ:

$$R = -\frac{4}{3}\pi a^3 \gamma - 2\pi a^3 \gamma \left[\frac{1}{6} - \frac{1}{2} \cos^2 \varphi \left(1 - \frac{2}{3} \cos \varphi \right) \right],$$ (7.41)

$$N_\varphi = \frac{\gamma a^2}{6} \left(5 + \frac{2 \cos^2 \varphi}{1 - \cos \varphi} \right),$$ (7.42)

$$N_\theta = \frac{\gamma a^2}{6} \left(1 - 6 \cos \varphi - \frac{2 \cos^2 \varphi}{1 - \cos \varphi} \right).$$ (7.43)

Using the expressions (7.39), (7.40), we can show that $N_\theta \geq 0$ and

$$N_\theta - N_\varphi = \frac{\gamma a^2}{6(1 + \cos \varphi)} \left(2 - \cos \varphi - \cos^2 \varphi \right) \geq 0$$ (7.44)

for the upper part of the shell ($0 \leq \varphi \leq \varphi_s$). Equality in (7.44) is realized when $\varphi = 0$. For the lower part of the shell ($\varphi_1 < \varphi \leq \pi$) the membrane forces, defined by the expressions (7.41)–(7.43), satisfy the inequalities $N_\varphi > 0$ and

$$N_\varphi - N_\theta = \frac{\gamma a^2}{3(1 - \cos \varphi)} \left(2 + \cos \varphi + \cos^2 \varphi \right) > 0.$$ (7.45)

Thus, for the upper part of the shell, the third term in (7.27) is larger than the second, and for the lower part of the shell the second term in (7.27) is larger than the third. Consequently, the optimal design of a spherical tank can be written as

$$h = \max\left\{ h_0, \frac{\gamma a^2 \sqrt{\pi l_m/2}}{6K_{1\varepsilon}} \left(5 - 6\cos\varphi + \frac{2\cos^2\varphi}{1 + \cos\varphi} \right) \right\}, \quad 0 \le \varphi < \varphi_s,$$

$$h = \max\left\{ h_0, \frac{\gamma a^2 \sqrt{\pi l_m/2}}{6K_{1\varepsilon}} \left(5 + \frac{2\cos^2\varphi}{1 - \cos\varphi} \right) \right\}, \quad \varphi_s \le \varphi < \pi. \tag{7.46}$$

7.3 Some Examples with Prescribed Middle Surfaces

The solution (7.27) was obtained using $r_\varphi = r_\varphi(\varphi)$, $r_\theta = r_\theta(\varphi)$ and $r = r(\varphi)$. But for many problems of optimization of axisymmetrical shells it is more convenient to take the coordinate x as an independent variable, to use the representation $h = h(x)$, $r = r(x)$ for desired design variables and to exclude the functions $r_\varphi(\varphi)$, $r_\theta(\varphi)$ from consideration with the help of relations (7.1), (7.2). We illustrate this approach below using some examples with prescribed middle surface. We note also that this formulation is very convenient not only for problems with unknown thickness distribution but also for the shape optimization problems.

Let us consider the shell with the length L that is described by the radius $r = r(x)$ ($0 \le x \le L$) and thickness distribution $h = h(x)$. Suppose that the shell is loaded by the constant internal pressure $q_n = q$ and distributed axisymmetric forces applied at its ends: $x = 0$ and $x = L$, the resultants of which R_1 ($x = 0$) and R_2 ($x = L$) are directed along the axis of the shell. The volume of the materials of the shell is given by the formula

$$J = 2\pi \int_0^L rh \sqrt{1 + \left(\frac{dr}{dx} \right)^2}\, dx \tag{7.47}$$

and the expressions for the normal stresses have the form

$$\sigma_\varphi = \frac{N_\varphi}{h} = \frac{R_1 + \pi q(r^2(x) - r^2(0))}{2\pi h(x) r(x)} \sqrt{1 + \left(\frac{dr}{dx} \right)^2}, \tag{7.48}$$

$$\sigma_\theta = \frac{N_\theta}{h} = \frac{q r(x)}{h(x)} \sqrt{1 + \left(\frac{dr}{dx} \right)^2} + \frac{R_1 + \pi q(r^2(x) - r^2(0))}{2\pi h(x) \sqrt{1 + \left(\frac{dr}{dx} \right)^2}} \left(\frac{d^2 r}{dx^2} \right). \tag{7.49}$$

The inequalities (7.24), (7.25) are written as

$$\max_{0 \le x \le L} \left(\sqrt{\pi l_m/2} \, \sigma_\varphi \right) \le K_{1\varepsilon}, \tag{7.50}$$

$$\max_{0 \le x \le L} \left(\sqrt{\pi l_m/2} \, \sigma_\theta \right) \le K_{1\varepsilon}. \tag{7.51}$$

If the problem parameter

$$\sigma_* = \frac{K_{1\varepsilon}}{\sqrt{\pi l_m/2}} \tag{7.52}$$

is introduced then the strength conditions (7.50), (7.51) can be written as

$$\max_{0 \le x \le L} \sigma_\varphi \le \sigma_*, \tag{7.53}$$

$$\max_{0 \le x \le L} \sigma_\theta \le \sigma_* \tag{7.54}$$

or in more compact form

$$\max \{\sigma_\varphi, \sigma_\theta\} \le \sigma_*, \ 0 \le x \le L. \tag{7.55}$$

The operation max in (7.55) selects the larger of the two quantities inside the braces.

The optimization problem consists in finding the thickness distribution $h = h(x)$ and/or radius $r = r(x)$, such that the cost functional (7.47) attains a minimum while satisfying the geometric constraint (7.18) and the strength constraint (7.55).

Suppose that $q = 0$, and consider some examples in which the middle surface of the shell is fixed, i.e. the function $r = r(x)$ is given, and the thickness distribution $h = h(x)$ is taken as an unknown design variable. The corresponding optimization problem consists in finding $h = h(x)$ such that the functional (7.47) attains a minimum under the constraints (7.55) and (7.18) with

$$\chi = h(x) - h_0 \ge 0. \tag{7.56}$$

Using the inequalities (7.56) and (7.55) and the expressions (7.47)–(7.49), we find the optimal design of the shell in the following form:

$$h = \max \left\{ h_0, \ \frac{N_\varphi}{\sigma_*}, \ \frac{N_\theta}{\sigma_*} \right\} =$$

$$= \max \left\{ h_0, \ \frac{R_1}{2\pi\sigma_* r} \sqrt{1 + \left(\frac{dr}{dx}\right)^2}, \ \frac{R_1}{2\pi\sigma_* \sqrt{1 + \left(\frac{dr}{dx}\right)^2}} \left(\frac{d^2 r}{dx^2}\right) \right\}. \tag{7.57}$$

The max operation in (7.57) finds the maximum of the three values for any fixed $x \in [0, L]$. To illustrate the solution (7.57), we consider the particular case when the shape of the middle surface of the shell is conical (see Fig. 7.6) and r is linear:

$$r(x) = a + \frac{b-a}{L}x = a + kx, \ 0 \le x \le L.$$

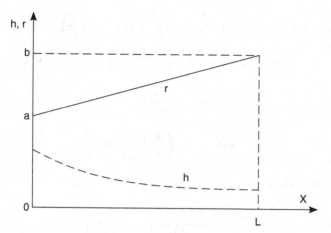

Fig. 7.6 The case of a conical shell

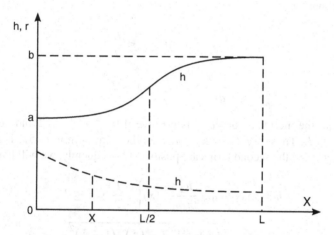

Fig. 7.7 A shell with a polynomial radius r

In this case, the optimal thickness distribution is given by the formula

$$h(x) = \max \left\{ h_0, \ \frac{R_1 \sqrt{1 + k^2}}{2\pi\sigma_*(1 + kx)} \right\} \tag{7.58}$$

and is shown in Fig. 7.6 by a dashed line. The continuous line represents the given radius distribution. The optimal design shown in Fig. 7.6 corresponds to the case when the constraint (7.56) is inactive, and the maximum is realized for the second term in (7.58).

As an other example illustrating the solution (7.57), we suppose the shape (see Fig. 7.7) of the middle surface is

$$r = r(x) = -2(b - a)\left(\frac{x}{L}\right)^3 + 3(b - a)\left(\frac{x}{L}\right)^2 + a. \tag{7.59}$$

The curve $r(x)$ satisfies the following boundary condition at $x = 0$ and $x = l$:

$$r(0) = a, \quad \left(\frac{dr}{dx}\right)_{x=0} = 0,$$

$$r(L) = b, \quad \left(\frac{dr}{dx}\right)_{x=L} = 0. \tag{7.60}$$

Taking into account that

$$\frac{d^2 r}{dx^2} = \frac{6(b - a)}{L^2}\left[1 - 2\left(\frac{x}{L}\right)\right] \tag{7.61}$$

and, consequently,

$$\frac{d^2 r}{dx^2} \geq 0, \ 0 \leq x \leq \frac{L}{2},$$

$$\frac{d^2 r}{dx^2} \leq 0, \ \frac{L}{2} \leq x \leq L, \tag{7.62}$$

we note that the meridian curvature is positive if $0 \leq x \leq L/2$ and negative if $L/2 \leq x \leq L$. Thus, if $L/2 \leq x \leq L$ then the third term in (7.57) is negative, while the first and the second terms are positive. Consequently, we will have

$$h = \max\left\{h_0, \frac{R_1}{2\pi\sigma_*}\Phi(x)\right\}, \ \frac{L}{2} \leq x \leq L,$$

$$\Phi(x) \equiv \frac{\sqrt{1 + \frac{36(b-a)^2}{L^2}\left(\frac{x}{L}\right)^2\left(1 - \frac{x}{L}\right)^2}}{a + 3(b - a)\left(\frac{x}{L}\right)^2 - 2(b - a)\left(\frac{x}{L}\right)^3}. \tag{7.63}$$

All three various possibilities in (7.57) can be realized, depending on the problem parameters h_0, a, b, L, l_m, $K_{1\varepsilon}$, R_1 for the part of the shell $0 \leq x \leq L/2$ with positive meridian curvature, i.e.,

$$h = \max\left\{h_0, \frac{R_1}{2\pi\sigma_*}\Phi(x), \frac{R_1}{2\pi\sigma_*}\Psi(x)\right\}, \ 0 \leq x \leq \frac{L}{2},$$

$$\Psi(x) \equiv \frac{\frac{6(b-a)}{L^2}\left[1 + 2\left(\frac{x}{L}\right)\right]}{\sqrt{1 + \frac{36(b-a)^2}{L^2}\left(\frac{x}{L}\right)^2\left(1 - \frac{x}{L}\right)^2}}. \tag{7.64}$$

If h_0 satisfies the inequality $R_1 \Phi(x)/2\pi\sigma_* \geq h_0$ for all $x \in [0, L]$ then the optimal thickness distribution of the shell takes the form

$$
h = \begin{cases} \frac{R_1}{2\pi\sigma_*} \Psi(x), & 0 \leq x \leq x_*, \\ \frac{R_1}{2\pi\sigma_*} \Phi(x), & x_* \leq x \leq L, \end{cases} \tag{7.65}
$$

where x_* is a unique solution of the equation $\Phi(x) - \Psi(x) = 0$ on the interval $(0, L)$. The dashed line in Fig. 7.7 shows the optimal design.

Chapter 8
Shape and Thickness Distribution
of Pressure Vessels

We consider simultaneous combined optimization of the shape and thickness of membrane shells of revolution under the action of internal pressure. We take account of the constraints concerning the strength of the shell and the volume of its cavity [Ban07]. We give general formulations of optimal design of closed shells of revolution (pressure vessels), and investigate the optimal shape of a shell and the corresponding thickness distribution. We present exact solutions for the optimal design of closed shells of revolution under internal pressure. The simultaneous introduction of two control functions, describing the shape of the shell and the distribution of its thickness not only ensures a substantial reduction in the mass of a shell, but also leads to significant mathematical simplifications: this leads to closed form solution of the optimization problems.

Note that many papers (see [TW59, BBS05], for example), which take a fixed shape of the neutral surface and a thickness distribution which is optimizable along a given surface, have been devoted to optimal design of axisymmetric shells and composite shells of revolution. The search for the optimal shape of the middle surface of a shell for a specified thickness distribution presents considerable difficulties even for the case of constant thickness, and leads to the problems which can be solved only numerically (see [SK89, ETC96]).

8.1 Optimizing the Shape of a Shell of Revolution

Consider a shell which has the shape of a surface of revolution, the axis of which coincides with the x-axis (Fig. 8.1). The position of the meridian plane is specified by the angle θ measured from a certain fixed plane. The alignment of the parallel circle is defined by the angle φ between the normal to the surface and the axis of rotation; $r = r(x)$ is the radius of the parallel circle, which determines the distance from a point on the neutral surface of the shell to its axis of rotation; x satisfies $0 \le x \le L$, L is the specified length of the shell. The quantities $r(0) = r_1$ and $r(L) = r_2$ are assumed to be given and to satisfy the inequalities $r_1 \ge 0$, $r_2 \ge 0$, $r_1 \le r_2$. The thickness distribution $h = h(x)$ and the radius $r = r(x)$ are assumed to satisfy the inequality (7.4), (7.5), written as

N.V. Banichuk and P.J. Neittaanmäki, *Structural Optimization with Uncertainties*, Solid
Mechanics and Its Applications 162, DOI 10.1007/978-90-481-2518-0_1,
© Springer Science+Business Media B.V. 2010

Fig. 8.1 Meridian plane of a
shell of revolution

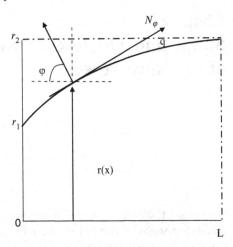

$$h(x) \le h_m = \max_x h(x) \ll r_m,$$

$$r_m = \min \left\{ \min_x r_\varphi(x), \min_x r_\theta(x) \right\}.$$

Here r_φ, r_θ are corresponding principal radii of curvature, and minima with respect to x refer to the interval $[0, L]$.

The shell is loaded with a constant internal pressure q, and by distributed forces on its ends: $x = 0$ and $x = L$, the resultants of which $R_1(x = 0)$ and $R_2(x = L)$ are directed along the axis of the shell. If the ends $x = 0$ and $x = L$ of the shell are fitted with circular end plates, then $R_1 = \pi r_1^2 q$, $R_2 = \pi r_2^2 q$. If the shell has poles $r(0) = r_1 = 0$ and $r(L) = r_2 = 0$ at its ends as, for example, for closed ellipsoidal shells of revolution then $R_1 = 0$, $R_2 = 0$.

The equation of equilibrium for the cut off part of the shell $x < L$ in the axial direction can be written as follows:

$$\frac{2\pi r}{\sqrt{1 + \left(\frac{dr}{dx}\right)^2}} N_\varphi = R_1 + \pi q \left(r^2 - r_1^2\right) \tag{8.1}$$

while the equilibrium equation of an element of a membrane shell, written for the direction normal to the neutral surface of the shell and which relates the magnitudes of normal membrane forces N_φ, N_θ has the form (7.6). The normal stresses arising in the shell are $\sigma_\varphi = N_\varphi/h$, $\sigma_\theta = N_\theta/h$, $\sigma_{\varphi\theta} = 0$. For a fixed shape of the middle surface of the shell $r = r(x)$ and a specified thickness distribution $h = h(x)$, the volume of the material of the shell is

$$J = J(r, h) = 2\pi \int_0^L rh \sqrt{1 + \left(\frac{dr}{dx}\right)^2} \, dx. \tag{8.2}$$

The optimization problem being considered consists of searching for the shape of the shell $r = r(x)$ and its thickness distribution $h = h(x)$ which make the volume of the shell material (8.2) a minimum when the strength condition

$$\max \{\sigma_\varphi, \sigma_\theta\} \le \sigma_* \tag{8.3}$$

the boundary conditions

$$r(0) = r_1, \quad r(L) = r_2 \tag{8.4}$$

and the geometrical constraint imposed on the volume of the shell cavity

$$\pi \int_0^L r^2 dx = V_0 \tag{8.5}$$

are satisfied, where σ_*, r_1, r_2, L and V_0 are specified positive constants. The strength condition can be written in the form of (8.3) for membrane shells of revolution made of brittle and quasibrittle materials. At the same time the constant of strength of the material σ_* (the reduced maximum stress) is determined using known values of the constant K_{1C} (toughness of quasibrittle material) and the maximum admissible length l_m of the cracks, which arise in the manufacture of a shell or as a result of its use [ETC96, KP85, Hut79].

8.2   The Search for the Shape and Thickness Distribution of an Optimal Shell

We will consider the problem of the optimal design of a closed axisymmetric shell with rigid end plates at $x = 0$ and $x = L$ that is acted upon by an internal pressure q. We have

$$R_1 = \pi r_1^2 q, \quad R_2 = \pi r_2^2 q. \tag{8.6}$$

When constructing the optimal solution, we shall assume that, in inequality (8.3), the sign of strict equality holds over the whole of the interval $x \in [0, L]$:

$$\sigma_\varphi = \frac{N_\varphi}{h} = \sigma_*, \quad \sigma_\theta \le \sigma_* \tag{8.7}$$

or

$$\sigma_\theta = \frac{N_\theta}{h} = \sigma_*, \quad \sigma_\varphi \le \sigma_*. \tag{8.8}$$

We shall initially assume that the conditions (8.7) are satisfied over the whole of the interval $[0, L]$, that is, the constraint imposed on the meridian stress σ_φ is active. It then follows from relations (8.1), (8.7) and (8.6) that the optimal thickness distribution $h = h_\varphi = h_\varphi(x)$ and the corresponding optimal shape of the shell $r = r(x)$ are related by the equality

$$h_\varphi = \frac{qr}{2\sigma_*} \sqrt{1 + \left(\frac{dr}{dx}\right)^2}. \tag{8.9}$$

The subscript φ on the distribution of the thickness h denotes that, when relation (8.9) is satisfied, the meridian stress σ_φ is critical and the peripheral stress σ_θ is assumed to satisfy the inequality $\sigma_\theta \le \sigma_\varphi$.

Using the relations (7.6), (8.1), (8.5), and (8.9), we arrive at the following representations for the minimized volume of the shell material (the functional of the problem) and for the Lagrange augmented functional, which takes account of the given volume of the shell cavity (8.5)

$$J = 2\pi \int_0^L rh_\varphi \sqrt{1 + \left(\frac{dr}{dx}\right)^2} \, dx = \frac{\pi q}{\sigma_*} \int_0^L r^2 \left(1 + \left(\frac{dr}{dx}\right)^2\right) dx, \tag{8.10}$$

$$J^a = J - \lambda V_0 = \frac{\pi q}{\sigma_*} \int_0^L \left\{\left(r\frac{dr}{dx}\right)^2 + \beta r^2\right\} dx, \quad \beta = 1 - \frac{\lambda\sigma_*}{q}, \tag{8.11}$$

where λ is the Lagrange multiplier corresponding to the condition (8.5). The necessary condition for an extremum of the functional (8.11) (Euler's equation) under the assumption that $r(x) \ne 0$ when $x \in [0, L]$ has the form

$$r\frac{d^2r}{dx^2} + \left(\frac{dr}{dx}\right)^2 - \beta = 0. \tag{8.12}$$

In this case, the search for the shape of the optimal shell reduces to solving the following boundary value problem:

$$\frac{d^2\left(r^2\right)}{dx^2} = 2\beta, \quad r(0) = r_1, \quad r(L) = r_2. \tag{8.13}$$

We now introduce the notation

$$\Delta_\pm = \left(r_2^2 \pm r_1^2\right)/L^2.$$

The solution of problem (8.13) has the form

$$r^2 = \beta x^2 + ax + r_1^2, \quad a = L(\Delta_- - \beta). \tag{8.14}$$

The optimal thickness distribution $h_\varphi = h_\varphi(x)$, corresponding to the optimal shape which has been found, is obtained using relations (8.9) and (8.14) and is written in the form

$$h_\varphi = \frac{q}{2\sigma_*} \sqrt{\beta(1 + \beta)x^2 + a(1 + \beta)x + a^2/4 + r_1^2}. \tag{8.15}$$

The Lagrange multiplier λ and the parameter β are found using expression (8.14), the isoperimetric condition (8.5) imposed on the volume of the shell cavity and formula (8.11) which relates the quantities λ and β. We have

$$\beta = -\alpha, \quad \alpha = \frac{6V_0}{\pi L^3} - 3\Delta_+,$$

$$\lambda = \frac{q}{\sigma_*}(1 - \beta) = \frac{q}{\sigma_*}(1 + \alpha). \tag{8.16}$$

It follows from the relations (8.10) and (8.14) that, in the case of the solution (8.14)–(8.16) being considered, the minimum volume of shell material is given by the formula

$$J = J_\varphi = \frac{q}{\sigma_*}V_0 + \frac{\pi q L}{\sigma_*}\left(\frac{\beta^2 L^2}{3} + \frac{\beta La}{2} + \frac{a^2}{4}\right) = \alpha(\alpha + 2) + 6\Delta_+ + 3\Delta_-^2. \tag{8.17}$$

The domain of variation of the parameters of the problem V_0, L, r_1, r_2, which ensures the existence of the solution of the form being considered, is established using the strength condition

$$\sigma_\theta(x) \leq \sigma_\varphi(x) = \sigma_*, \quad 0 \leq x \leq L,$$

which we convert to the inequality

$$\sigma_\theta = \frac{\sigma_*\left(2 + \left(\frac{dr}{dx}\right)^2 + \beta\right)}{1 + \left(\frac{dr}{dx}\right)^2} \leq \sigma_*, \quad 0 \leq x \leq L.$$

The solution of this inequality leads to the constraint

$$\alpha = -\beta = \frac{6V_0}{\pi L^3} - 3\Delta_+ \geq 1. \tag{8.18}$$

The optimal shapes of the shells $r(x)$ and their thickness distribution $h = h_\varphi(x)$ which have been found are shown in Fig. 8.2 for $\alpha = 5, 2, 1$ in the case when

$$r(0) = \frac{L}{2}, \quad r(L) = L. \tag{8.19}$$

It can be seen from relations (8.14) and (8.15) and the graphs that the optimal shell when $\alpha \geq 1$ has the shape of the part of an ellipsoid of revolution located between the cross-section $x = 0$ and $x = L$. The corresponding thickness distribution has it maximum values close to the ends of the shell and, when $\alpha \rightarrow 1$, the thickness of the shell tends to a constant value

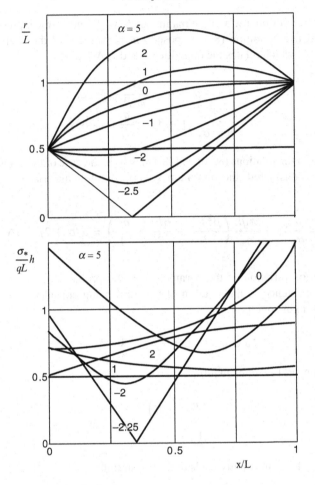

Fig. 8.2 Shape and thickness distributions

$$h_\varphi(x) = \frac{q}{2\sigma_*} \sqrt{\frac{a^2}{4} + r_1^2}.$$

When $r(0) = r(L)$, the shape of the shell and the thickness distribution are symmetrical about the cross-section $x = L/2$. If $r(0) = r(L) = 0$, the optimal shell takes the shape of an oblate ellipsoid of revolution

$$\tilde{r}^2 + \alpha \left(\tilde{x} - \frac{1}{2}\right)^2 = \frac{\alpha}{4}$$

$$\tilde{r} = \frac{r}{4}, \quad \tilde{x} = \frac{x}{L}$$

and the maximum value of the thickness, shown in Fig. 8.3, is attained at the poles (ends) of the ellipsoid.

Fig. 8.3 Thickness distribution

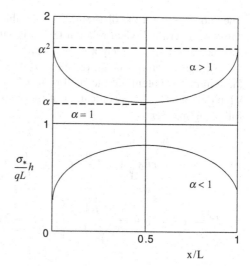

Fig. 8.4 Dependence of the shell mass on the problem parameter α

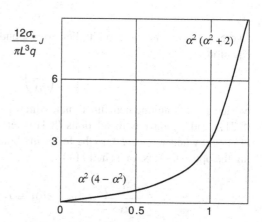

When $\alpha = 1$, the optimal shell acquires a spherical shape and its thickness becomes constant. The dependence of the mass of the shell on the parameter α when $\alpha \geq 1$ and $r(0) = r(L) = 0$ is represented in dimensionless form in Fig. 8.4.

We will now consider another case when conditions (8.8) are satisfied over the whole of the interval $[0, L]$, that is, the constraint imposed on the peripheral stress σ_θ is active. According to relations (7.6), (8.1), (8.6) and (8.8), the optimal thickness distribution $h = h_\theta = h_\theta(x)$ and the corresponding optimal shape of the shell $r = r(x)$ are connected by the relation

$$h_\theta = \frac{qr}{2\sigma_*} \frac{\left(\frac{1}{2}r\frac{d^2r}{dx^2} + \left(\frac{dr}{dx}\right)^2 + 1\right)}{\sqrt{1 + \left(\frac{dr}{dx}\right)^2}}. \tag{8.20}$$

The subscript θ on the distribution of the thickness h denoted that the peripheral stress σ_θ is critical when relation (8.19) is satisfied and the meridian stress σ_φ is assumed to satisfy the inequality $\sigma_\varphi \leq \sigma_\theta$.

Using the relations (8.2), (8.5) and (8.20) we can represent the expression for the minimized volume of the shell material by an augmented Lagrange functional, which allows for the specification of the volume of the cavity of the shell, in the following manner:

$$J = 2\pi \int_0^L r h_0 \sqrt{1 + \left(\frac{dr}{dx}\right)^2}\, dx = \frac{\pi q}{\sigma_*} \int_0^L r^2 \left[r\frac{d^2 r}{dx^2} + 2\left(\frac{dr}{dx}\right)^2 + 2 \right] dx,$$

$$J^a = J - \lambda V_0 = \frac{\pi q}{\sigma_*} \int_0^L \left\{ r^2 \left[r\frac{d^2 r}{dx^2} + 2\left(\frac{dr}{dx}\right)^2 \right] - \gamma r^2 \right\} dx, \qquad (8.21)$$

$$\gamma = \frac{\lambda \sigma_*}{q} - 2,$$

where λ is the Lagrange multiplier corresponding to condition (8.5). The Euler equation

$$r\frac{d^2 r}{dx^2} + \left(\frac{dr}{dx}\right)^2 - \gamma = 0 \qquad (8.22)$$

written for the augmented functional, which is defined by the second equality of (8.21), leads, jointly with relations (8.4) when account is taken of the inequality $r(x) \neq 0$ when $x \in (0, L)$, to the following boundary-value problem for determining the optimal shape of a shell $r(x)$

$$\frac{d^2(r^2)}{dx^2} = 2\gamma, \quad r(0) = r_1, \quad r(L) = r_2 \qquad (8.23)$$

The optimal solution, which is found from relations (8.20), (8.22) and (8.23), has the form

$$r^2 = \gamma x^2 + bx + r_1^2, \quad b = L(\Delta_- - \gamma), \qquad (8.24)$$

$$h_\theta = \frac{q}{2\sigma_*} \left[\chi + (1 + \gamma)\left(\gamma x^2 + bx + r_1^2\right) \chi^{-1} \right],$$

$$\chi = \sqrt{\gamma(1 + \gamma) x^2 + (1 + \gamma) bx + b^2 / 4 + r_1^2}. \qquad (8.25)$$

The parameters γ and λ are defined as

$$\gamma = -\frac{12 V_0}{\pi L^3} + 6\Delta_+, \qquad (8.26)$$

$$\lambda = \frac{2q}{\sigma_*}\left(1 - \frac{3V_0}{\pi L^3} + \frac{3}{2}\Delta_+\right). \tag{8.27}$$

For the volume of the shell material corresponding to the given solution, we have the expression

$$J = J_\theta = \frac{q}{\sigma_*}(2 + \gamma)V_0 + \frac{\pi qL}{4\sigma_*}\left(\frac{\gamma^2 L^2}{3} + \frac{\gamma bl}{2} + \frac{b^2}{4}\right) =$$

$$= \alpha(4 - \alpha) + 6\Delta_+(2 - \alpha) + 3\frac{\Delta_-}{L^2}. \tag{8.28}$$

The domain of variation of the parameters which ensures the existence of the optimal solution in the form of (8.24)–(8.28) is established using the condition

$$\sigma_\varphi \leq \sigma_\theta = \sigma_*, \quad 0 \leq x \leq L.$$

The inequality $\sigma_\varphi \leq \sigma_\theta$ leads to the relation

$$r\frac{d^2 r}{dx^2} + \left(\frac{dr}{dx}\right)^2 + 1 \geq 0, \tag{8.29}$$

and the condition $\sigma_\varphi \leq \sigma_*$, using the relations (8.1), (7.8) and (8.19), is written in the form

$$\sigma_\varphi = \sigma_*\frac{1 + \left(\frac{dr}{dx}\right)^2}{r\frac{d^2 r}{dx^2} + 2\left(\frac{dr}{dx}\right)^2 + 2} \leq \sigma_*. \tag{8.30}$$

It is easily noted, using the inequality (8.29), that the denominator in the relation (8.30) is positive, and the following constraint, imposed on the parameters of the problem therefore arises from the inequality (8.30):

$$\alpha = -\gamma = \frac{6V_0}{\pi L^3} - 3\Delta_+ \leq 1. \tag{8.31}$$

The second constraint follows from the condition

$$r(x) \geq 0, \quad 0 \leq x \leq L, \tag{8.32}$$

and has the form

$$\alpha \geq \alpha_0, \quad \alpha_0 = -\frac{(r_1 + r_2)^2}{L^2}. \tag{8.33}$$

The curvature of the optimal shell in the direction of the meridian vanishes, that is,

$$|r_\varphi| = \infty, \quad 0 \leq x \leq L$$

for values of the parameter α which are determined from the condition

$$\frac{d^2 r}{dx^2} = 0, \quad 0 \le x \le L$$

and are equal to

$$\alpha_1 = -\frac{(r_1 - r_2)^2}{L^2}.$$

When $\alpha = \alpha_1$ the optimal shell has conical shape

$$r(x) = r_1 + x \frac{r_2 - r_1}{L}, \quad 0 \le x \le 1 \tag{8.34}$$

and when $\alpha = \alpha_0$, it is described by the two conical surfaces

$$r = -\frac{r_1 + r_2}{L} x + r_1, \quad \text{when} \quad 0 \le x \le x_0,$$

$$r = \frac{r_1 + r_2}{L} x - r_1, \quad \text{when} \quad x_0 \le x \le L_0, \tag{8.35}$$

$$x_0 = \frac{r_1}{r_1 + r_2} L$$

corresponding to the broken line in the upper part of Fig. 8.2. We note that the shape (8.34) is the limiting shape for the solutions being considered when $\alpha_0 \le \alpha \le 1$.

The shell of conical form (8.34) partitions the whole family of the optimal shell shapes which have been constructed into biconvex shells of positive Gaussian curvature

$$k = \frac{1}{r_\varphi r_\theta} > 0$$

when $\infty > \alpha > \alpha_1$, and convex-concave shells of negative Gaussian curvature ($k < 0$) when $\alpha_0 < \alpha < \alpha_1$. The optimal shapes of the shells $r = r(x)$ and their thickness distributions $h = h_\theta(x)$ in the case when the equalities (8.19) are satisfied are shown in Fig. 8.2 for $\alpha = 1, 0, -1, -2$ and -2.25. The optimal shape and thickness distribution, corresponding to the limiting value $\alpha = 1$ are identical to the corresponding curves obtained from the relations (8.14) and (8.15) when $\alpha = 1$. It can be seen from the upper part of Fig. 8.2 that, as α increases, the curvature of the shell changes sign.

When $r(0) = r(L)$, the optimal shape of the shell and the corresponding thickness distribution are symmetrical with respect to the transverse cross-section $x = L/2$. If $r(0) = r(L) = 0$, then the dependence of the mass of the shell on the dimensionless parameter α is given by the expression

$$\frac{12\sigma_*}{\pi q L^3} J = \alpha^2 \left(4 - \alpha^2\right)$$

which is shown in Fig. 8.4 for $\alpha \leq 1$. The optimal shell corresponding to this case has the shape of a prolate ellipsoid of revolution and the maximum values of the thickness are attained at the equator $(x = L/2)$.

8.3 Some Properties of the Optimal Solution

It was assumed when constructing the optimal solution that in the relation (8.3), strict equality holds over the whole of the interval $[0, L]$, that is, the relations (8.7) and (8.8) are satisfied. It will be proved below that, if a solution exist which satisfies these relations (the strict attainments of the limit values), then it will be the optimal solution.

We will prove this assertion by an indirect method. We will assume that the optimal solution $\left(r_{opt}(x), h_{opt}(x)\right)$ satisfies the strict inequality in (8.3), or

$$h_{opt} > \frac{N_\varphi}{\sigma_*}, \quad h_{opt} > \frac{N_\theta}{\sigma_*} \tag{8.36}$$

in a certain segment $[x_1, x_2]$ $(0 \leq x_1 < x_2 \leq L)$. The optimal solution $(r_{opt}(x), h_{opt}(x))$ which is being considered is assumed to satisfy conditions (8.7) and (8.8) in the remaining segments $0 \leq x \leq x_1$ and $x_2 < x < L$ of the interval $[0, L]$ that is, it is assumed that strict equality holds in relation (8.3). In this case, it is possible to construct an admissible design $\left(\widehat{r}(x), \widehat{h}(x)\right)$ in the following form:

$$\widehat{r} = r_{opt}, \quad 0 \leq x \leq l,$$

$$\widehat{h} = h_{opt} \quad \text{if} \quad 0 \leq x < x_1, \quad x_2 < x \leq L, \tag{8.37}$$

$$\widehat{h} = \frac{N_\varphi\left(r_{opt}\right)}{\sigma_*} < h_{opt} \quad \text{or} \quad \widehat{h} = \frac{N_\theta\left(r_{opt}\right)}{\sigma_*} < h_{opt} \quad \text{if} \quad x_1 \leq x \leq x_2.$$

The design $\left(\widehat{r}, \widehat{h}\right)$ is admissible, since the distributions $\widehat{r}(x), \widehat{h}(x)$ which have been presented satisfy the strength condition (8.3) and the isoperimetric constraint imposed on the volume of the shell cavity (8.5). Note that the admissible design (8.37) which has been presented satisfies the relation (8.3) strictly over the whole of the interval $[0, L]$. Hence, the admissible solution constructed is an equal strength design, and for it we shall have

$$J(\widehat{r}, \widehat{h}) = I_{0,x_1} + 2\pi \int_{x_1}^{x_2} r_{opt}\widehat{h}\sqrt{1 + \left(\frac{dr_{opt}}{dx}\right)^2} \, dx + I_{x_2,L} < I_{0,L} = J\left(r_{opt}, h_{opt}\right),$$

$$\tag{8.38}$$

where

$$I_{a,b} = 2\pi \int_a^b r_{opt} h_{opt} \sqrt{1 + \left(\frac{dr_{opt}}{dx}\right)^2} \, dx.$$

The inverse equality established in this manner

$$J(r_{opt}, h_{opt}) > J\left(\widehat{r}, \widehat{h}\right)$$

proves the assertion that the relation (8.3) with strict equality holds over the whole of the interval $[0, L]$ when there is an optimal solution.

Thus when $\alpha \geq 1$, the shape of the optimal shell is specified by a biconvex surface of positive Gaussian curvature and, in particular, if the radii of the shell at its ends are equal to zero, then the shell has the shape of an oblate ellipsoid of revolution. A shell of spherical shape corresponds to the value of the dimensionless parameter $\alpha = 1$. If $\alpha = 1$, the type of shell changes and, when $\alpha < 1$, the peripheral stress σ_θ is critical for the shapes which have been found. However, when α is reduced, the optimal shell continues to have a positive Gaussian curvature up to a value $\alpha = \alpha_1$. When $\alpha = \alpha_1$ the curvature becomes zero and the optimal shell takes a conical shape. Then, when $\alpha_1 > \alpha > \alpha_0$, the optimal shell takes the shape of a convex-concave surface of revolution of negative Gaussian curvature. When $\alpha \to \alpha_0$, the surface of the optimal shell tends to a surface composed of conical surface and, moreover, at the point $x = x_0$, the radius of the shell $r(x_0) = 0$. The construction of the optimal shapes for the problem in question when $\alpha < \alpha_0$ leads to the search for solutions which have segments where $r = 0$, and they are omitted as not being of any physical interest.

Chapter 9
Brittle and Quasi-Brittle Materials

The optimization problems considered in this section consist in finding the shape and the thickness distribution of axisymmetric shells in the framework of membrane theory, loaded by fixed statical forces in such a way that the cost functional reaches a maximum, while satisfying some strength mechanics constraints. In this section the mass effectiveness of the shell is considered as a cost functional, and as constraints we use bounds on the maximum normal stresses; this is typical for shells made from brittle or quasi-brittle materials. Analytical investigations and corresponding optimal solutions are presented [BRS08].

9.1 Mass Effectiveness and Its Maximization

In this section we consider an elastic shell that has the form of a surface of revolution. The analysis presented is in the framework of membrane theory; this means that results are not valid in small regions surrounding possible kinks. Let the axis of shell symmetry coincide with the axis x. The position of the meridian plane is defined by the angle ϑ, measured from a datum meridian plane, and the position of a parallel circle is defined by the angle φ, made by the normal to surface and the axis of rotation, or alternatively by the coordinate x, measured along the axis of rotation. In what follows it is convenient to describe the geometry of the shell by a distance $r = r(x)$ from the rotation axis to a point of the middle surface (radius of the shell) and by a thickness distribution $h = h(x)$, $0 \leq x \leq L$, where L is the length of the shell (see Fig. 9.1).

It is assumed that the shell is subjected to pressure loads, and these loads act normal to the middle surface. The mass M of the shell material and the volume V of the shell are given by

$$M = 2\pi\rho \int_0^L rh \sqrt{1 + \left(\frac{dr}{dx}\right)^2}\, dx, \quad V = \pi \int_0^L r^2 dx. \tag{9.1}$$

N.V. Banichuk and P.J. Neittaanmäki, *Structural Optimization with Uncertainties*, Solid Mechanics and Its Applications 162, DOI 10.1007/978-90-481-2518-0_1,

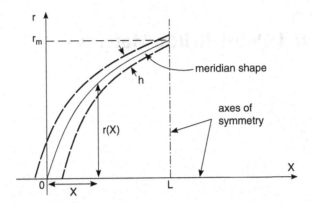

Fig. 9.1 A half of axisymmetric shell

The magnitude N_φ and N_θ of the membrane forces per unit length, and the normal membrane stresses σ_φ and σ_θ acting in the meridional and circumferential directions, are found with the help of equilibrium equation [Flu73, Tim56, TW59]. We have

$$\sigma_\varphi = \frac{N_\varphi}{h} = \frac{qr}{2h}\sqrt{1 + \left(\frac{dr}{dx}\right)^2}, \tag{9.2}$$

$$\sigma_\theta = \frac{qr}{2h}\left(\frac{2 + 2\left(\frac{dr}{dx}\right)^2 + r\frac{d^2r}{dx^2}}{\sqrt{1 + \left(\frac{dr}{dx}\right)^2}}\right). \tag{9.3}$$

The optimization problem consists in finding the shape $r = r(x)$ and the thickness distribution $h = h(x)$ of the shell such that the functional of mass effectiveness

$$K = \frac{qV}{M} = \frac{q}{2\rho}\frac{\int_0^L r^2 dx}{\int_0^L rh\sqrt{1 + \left(\frac{dr}{dx}\right)^2}dx} \tag{9.4}$$

is maximized while satisfying the strength constraint

$$\max\{\sigma_\varphi, \sigma_\theta\} \le \sigma_* \tag{9.5}$$

and the boundary conditions for $r(x)$:

$$r(0) = 0, \quad r(L) = r_m, \tag{9.6}$$

where the length L, the right radius r_m and the strength constant σ_* of the shell material are given positive parameters.

In what follows we will solve the optimization problem using the equality $\sigma_\varphi = \sigma_*$, where σ_φ is given by (9.2). Let us prove that the inequality (9.5) must be satisfied strictly. Suppose that

$$\sigma_\theta \leq \sigma_\varphi \leq \sigma_*.$$

In this case, as is seen from (9.2) and (9.5), all admissible thickness distributions h_{ad} must satisfy the inequality

$$h_{ad} \geq h_\varphi = \frac{qr}{2\sigma_*}\sqrt{1+\left(\frac{dr}{dx}\right)^2} \tag{9.7}$$

for arbitrary continuous function $r = r(x)$ satisfying given boundary conditions. Consequently we will have

$$K(r,h_{ad}) = \frac{q}{2\rho}\frac{\int\limits_0^L r^2 dx}{\int\limits_0^L rh_{ad}\sqrt{1+\left(\frac{dr}{dx}\right)^2}dx} \leq \frac{q}{2\rho}\frac{\int\limits_0^L r^2 dx}{\int\limits_0^L rh_\varphi\sqrt{1+\left(\frac{dr}{dx}\right)^2}dx} = K(r,h_\varphi) \tag{9.8}$$

and, therefore,

$$J(r,h_{ad}) = \frac{\rho}{\sigma_*}K(r,h_{ad}) \leq \frac{\int\limits_0^L r^2 dx}{\int\limits_0^L r^2\left(1+\left(\frac{dr}{dx}\right)^2\right)dx} = J_\varphi(r). \tag{9.9}$$

In the same manner we can consider the case

$$\sigma_\varphi \leq \sigma_\theta \leq \sigma_*$$

and show that for

$$h_{ad} \geq h_\theta = \frac{qr}{2\sigma_*}\frac{\left(r\frac{d^2r}{dx^2}+2\left(\frac{dr}{dx}\right)^2+2\right)}{\sqrt{1+\left(\frac{dr}{dx}\right)^2}} \tag{9.10}$$

the following inequality is satisfied:

$$J(r,h_{ad}) = \frac{\rho}{\sigma_*}K(r,h_{ad}) \leq \frac{\rho}{\sigma_*}K(r,h_\theta) = \frac{\int\limits_0^L r^2 dx}{\int\limits_0^L r^2\left(r\frac{d^2r}{dx^2}+2\left(\frac{dr}{dx}\right)^2+2\right)dx} = J_\theta(r). \tag{9.11}$$

9.2 Analytical Solution and Convexity Properties

Suppose that the maximum in (9.5) is realized for the first value; the equality $\sigma_\varphi = \sigma_*$ and the strength constraint (9.5) imply:

$$\sigma_\theta \leq \sigma_\varphi = \sigma_*, \tag{9.12}$$

$$h = h_\varphi = \frac{qr}{2\sigma_*}\sqrt{1 + \left(\frac{dr}{dx}\right)^2}. \tag{9.13}$$

It is convenient to exclude the design variable $h(x)$ from the expressions (9.1) and (9.4) by using the relation (9.13). We obtain the following representation:

$$M = \frac{\pi\rho q}{\sigma_*}J_M, \tag{9.14}$$

$$J_M = \int_0^L r^2\left(1 + \left(\frac{dr}{dx}\right)^2\right)dx, \tag{9.15}$$

$$V = \pi J_V, \tag{9.16}$$

$$J_V = \int_0^L r^2 dx, \tag{9.17}$$

$$J = J_\varphi = \frac{\rho}{\sigma_*}K = \frac{J_V}{J_M} \to \max_{r(x)}. \tag{9.18}$$

The optimization problem (9.6), (9.12)–(9.18) consists in maximizing the quality functional $J_\varphi = J_V/J_M$ and finding the extremum function $r = r(x)$ satisfying boundary conditions (9.6). The corresponding extremum thickness distribution is determined by formula (9.13). The inequalities (9.12) are used to derive the constraints for the problem parameters.

To obtain necessary optimality conditions, we perform the variation of the non-additive functional J_φ [CB73, Ban83, Ban90]. We have

$$\delta J_\varphi = \frac{1}{J_M}(\delta J_V - \lambda \delta J_M), \tag{9.19}$$

$$\lambda = \frac{J_V}{J_M} = J_\varphi. \tag{9.20}$$

Applying the necessary extremum condition

$$\delta J_\varphi = 0$$

and elementary transformations give us the following Euler equation:

$$\frac{d^2(r^2)}{dx^2} = \beta, \quad \beta = 2\left(\frac{\lambda - 1}{\lambda}\right). \tag{9.21}$$

To derive (9.21) we use also the condition that $r(x) > 0$ for $0 < x < L$.

Taking into account the expressions (9.20) for λ and (9.21) for β, we find the following constraints:

$$0 < \lambda < 1, \quad \beta < 0. \tag{9.22}$$

After integration of (9.21) with boundary conditions (9.6) we obtain (Fig. 9.2)

$$r^2(x) = \frac{\beta}{2}x(x - L) + \frac{x}{L}r_m^2, \quad 0 < x < L. \tag{9.23}$$

To compute β and, consequently, $\lambda(\lambda = 2/(2 - \beta))$, we will use the expression (9.23) for $r(x)$, the inequalities (9.22) and the following relation:

$$\lambda = \frac{J_V}{J_M} = \frac{J_V}{J_V + \int\limits_0^L r^2 \left(\frac{dr}{dx}\right)^2 dx}. \tag{9.24}$$

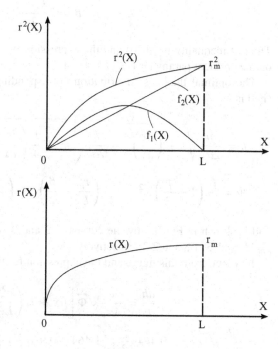

Fig. 9.2 Finding the shape

As a result we will have

$$\beta = 2\sqrt{3}\left(\sqrt{3} - 2\right)\left(\frac{r_m}{L}\right)^2 \approx \beta_0 \left(\frac{r_m}{L}\right)^2,$$

$$\beta_0 = -0.928, \tag{9.25}$$

$$\lambda = \frac{2}{2 - \beta} = \frac{1}{1 + 0.464\left(\frac{r_m}{L}\right)^2}.$$

The constructed solution is correct if the left inequality (9.12) is satisfied. Using (9.12) and the expressions (9.2) and (9.3) for σ_φ, σ_θ, we obtain the inequality

$$1 + \left(\frac{dr}{dx}\right)^2 + r\frac{d^2r}{dx^2} \leq 0.$$

When (9.21) is written in the form

$$r\frac{d^2r}{dx^2} + \left(\frac{dr}{dx}\right)^2 = \frac{\lambda - 1}{\lambda}$$

the inequality is transformed to $2 - \lambda^{-1} \leq 0$.

Consequently, we have

$$\lambda \leq \frac{1}{2}, \quad \beta = 2\frac{\lambda - 1}{\lambda} \leq -2. \tag{9.26}$$

The last inequality in (9.26) and the expression (9.25) for β give us the constraint on the problem parameter $(r_m/L) \geq 1.468$.

The optimal thickness distribution corresponding to the optimal shape of the shell is

$$h = \frac{qr}{2\sigma_*}\sqrt{1 + \left(\frac{dr}{dx}\right)^2} = \frac{q}{2\sigma_*}\sqrt{r^2 + \frac{1}{4}\left(\frac{dr^2}{dx}\right)^2} = \frac{q}{2\sigma_*}\sqrt{\Phi}, \tag{9.27}$$

$$\Phi = \frac{\beta}{2}\left(1 + \frac{\beta}{2}\right)x^2 + \frac{L^2}{4}\left(\frac{r_m^2}{L^2} - \frac{\beta}{2}\right)^2 + \left(1 + \frac{\beta}{2}\right)\left(\frac{r_m^2}{L^2} - \frac{\beta}{2}\right)Lx \tag{9.28}$$

and is shown in Fig. 9.3 by the curves 1, 2 and 3 for $q/2\sigma_* = 0.01$, $L = 1$ and $r_m/L = 1.5$, 1.8 and 2.2 respectively.

It is seen from this figure and the expression for the derivative, namely

$$\frac{dh}{dx} = \frac{q}{4\sigma_*}\sqrt{\Phi}\left[\beta x + L\left(\frac{r_m^2}{L^2} - \frac{\beta}{2}\right)\right]$$

$$= \frac{qL}{4\sigma_*}\sqrt{\Phi}\left(1 - 0.464\frac{r_m^2}{L^2}\right)\left(1.464 - 0.928\frac{x}{L}\right)\frac{r_m^2}{L^2} \leq 0, \quad 0 \leq x \leq L, \tag{9.29}$$

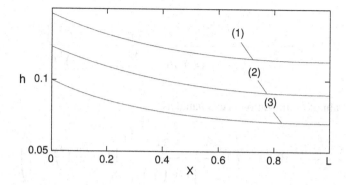

Fig. 9.3 Optimal thickness distribution corresponding to the obtained optimal shape of the shell

that the thickness distribution is a monotonically decreasing function on the interval $[0, L]$. The derivative dh/dx becomes positive when $x > x_*$, where x_* is the root of the equation $dh/dx = 0$, i.e.

$$x_* = -\frac{L}{\beta}\left(\frac{r_m^2}{L^2} - \frac{\beta}{2}\right) \geq \frac{4}{3}L. \qquad (9.30)$$

Let us compare the mass effectiveness

$$K_{opt} = \frac{\sigma_*}{\rho}\left(J_\varphi\right)_{opt} = \frac{\sigma_*}{\rho}\left[1 - \frac{\beta_0}{2}\left(\frac{r_m}{L}\right)^2\right] \qquad (9.31)$$

of this shell having optimal shape $r = r(x)$ and thickness distribution $h = h(x)$ with a part of a spherical shell having constant thickness, the same length L and radius $r = r_m$ at the point $x = L$. Note that the spherical shell is uniquely determined if parameters r_m and L are given, and that

$$r^2(x) = R^2 - (R - x)^2,$$
$$R = \frac{L}{2}\left[1 + \left(\frac{r_m}{L}\right)^2\right], \qquad (9.32)$$

where R is the radius of the spherical surface. To satisfy the strength constraint (9.5), the minimal thickness is determined as

$$h = \frac{q}{2\sigma_*}\sqrt{r^2 + \frac{1}{4}\left(\frac{dr^2}{dx}\right)^2} = \frac{qR}{2\sigma_*}. \qquad (9.33)$$

Using (9.32) and (9.33), we will obtain the following formulas for the mass effectiveness of the spherical shell:

$$K_S = \frac{q}{2\rho} \frac{\int\limits_0^L r^2 \, dx}{\int\limits_0^L rh\sqrt{1 + \left(\frac{dr}{dx}\right)^2} \, dx} = \frac{2\sigma_*}{3\rho} \frac{\left[1 + 3\left(\frac{r_m}{L}\right)^2\right]}{\left[1 + \left(\frac{r_m}{L}\right)^2\right]^2}. \qquad (9.34)$$

Thus the gain of optimization is evaluated as

$$\kappa = \frac{K_{opt} - K_S}{K_{opt}} 100\% = \left(1 - \frac{2}{3} \frac{\left[1 + 3\left(\frac{r_m}{L}\right)^2\right]}{\left[1 + \left(\frac{r_m}{L}\right)^2\right]^2} \left[1 - \frac{\beta_0}{2}\left(\frac{r_m}{L}\right)^2\right]\right) 100\%. \qquad (9.35)$$

For $r_m/L = 1.5$, 1.8 and 2.2 we have $\kappa = 0.02$, 0.48 and 1.53 respectively.

In general, the solution proposed has a discontinuity in the derivative of the shape function $r(x)$ (kink) for $x = L$. But if instead of the boundary conditions (9.6), we consider r_m as a variable parameter and adopt the boundary conditions

$$r(0) = 0, \quad \left(\frac{dr}{dx}\right)_{x=L} = 0$$

for $r(x)$ the optimal solution is an ellipsoid; this avoids the kink for $x = L$.

Suppose now that the maximum in (9.5) is realized for the second value. In this case, using the equality $\sigma_\theta = \sigma_*$ and the strength constraint (9.5), we obtain

$$\sigma_\varphi \leq \sigma_\theta \leq \sigma_*, \qquad (9.36)$$

$$h = h_\theta = \frac{qr}{2\sigma_*} \frac{\left(r\frac{d^2r}{dx^2} + 2\left(\frac{dr}{dx}\right)^2 + 2\right)}{\sqrt{1 + \left(\frac{dr}{dx}\right)^2}}. \qquad (9.37)$$

Let us use the relation (9.37) between h_θ and r and insert the corresponding expression into the basic relation of the optimization problem instead of h. As a result we obtain the following formulas:

$$J_M = \int\limits_0^L r^2 \left(2 + 2\left(\frac{dr}{dx}\right)^2 + r\frac{d^2r}{dx^2}\right) dx, \qquad (9.38)$$

$$J = J_\theta = \frac{\rho}{\sigma_*} K = \frac{J_V}{J_M} = \frac{\int\limits_0^L r^2 \, dx}{\int\limits_0^L r^2 \left(2 + 2\left(\frac{dr}{dx}\right)^2 + r\frac{d^2r}{dx^2}\right) dx} \to \max_{r(x)}. \qquad (9.39)$$

The necessary optimality condition (Euler equation) for the problem of maximization of the functional (9.39) and solution of this equation with the boundary conditions (9.6) can be written as

$$\frac{d^2(r^2)}{dx^2} = \gamma, \quad \gamma = \frac{2(1-2\lambda)}{\lambda}, \tag{9.40}$$

$$r^2(x) = \frac{\gamma}{2}x(x-L) + \frac{x}{L}r_m^2, \quad 0 < x < L. \tag{9.41}$$

The parameter γ in (9.40) and (9.41) must be determined with the help of the equation $\lambda = J_V/J_M$, which is transformed to the form

$$\gamma^2 = -12\left(\frac{r_m}{L}\right)^4.$$

The quantity γ is imaginary: the problem (9.39) and (9.6) has no solution.

We note some aspect of the optimization problem (9.6), (9.14)–(9.18) and study the convexity properties of the functionals. To do this, it is convenient for the sake of brevity to use the expression

$$I(\Psi) = \int_0^L \Psi \, dx.$$

Then we can write the maximized functional in the following manner:

$$J = J_\varphi = \frac{I\left(r^2\right)}{I\left(r^2\right) + I\left(r\left(\frac{dr}{dx}\right)^2\right)} = \frac{1}{1+\frac{T}{4}}, \tag{9.42}$$

where

$$T = \frac{I\left(\left(\frac{dr^2}{dx}\right)^2\right)}{I\left(r^2\right)}. \tag{9.43}$$

Maximizing the functional J_φ is equivalent to minimizing the functional T. We now use the new variable

$$u = r^2 \geq 0, \quad 0 \leq x \leq L. \tag{9.44}$$

Thus we have

$$T = \frac{I\left(\left(\frac{du}{dx}\right)^2\right)}{I(u)} \underset{u(x)}{\to} \min. \tag{9.45}$$

In what follow we prove that this problem is convex, i.e.

$$T(\alpha u_1 + (1-\alpha)u_2) \leq \alpha T(u_1) + (1-\alpha)T(u_2) \tag{9.46}$$

for any admissible functions

$$u_1 = r_1{}^2(x), \quad u_2 = r_2{}^2(x)$$

and for any α satisfying the inequalities $0 \leq \alpha \leq 1$.

Let us introduce a positive constant $\mu > 0$ such that

$$I(u_2) = \mu I(u_1). \tag{9.47}$$

Then we will use the quadratic inequality $2ab \leq a^2 + b^2$ to give

$$2 \frac{du_1}{dx} \frac{du_2}{dx} = 2 \left(\sqrt{\mu} \frac{du_1}{dx} \right) \left(\frac{1}{\sqrt{\mu}} \frac{du_2}{dx} \right) \leq \mu \left(\frac{du_1}{dx} \right)^2 + \frac{1}{\mu} \left(\frac{du_2}{dx} \right)^2 \tag{9.48}$$

and perform the following estimates:

$$T\left(\alpha u_1 + (1-\alpha)\,u_2\right) =$$

$$= \frac{\alpha^2 I \left(\left(\frac{du_1}{dx} \right)^2 \right) + (1-\alpha)^2 I \left(\left(\frac{du_2}{dx} \right)^2 \right) + 2\alpha(1-\alpha) I \left(\frac{du_1}{dx} \frac{du_2}{dx} \right)}{\alpha I(u_1) + (1-\alpha)\,I(u_2)}$$

$$\leq \frac{\alpha \left[\alpha + (1-\alpha)\,\mu \right] I \left(\left(\frac{du_1}{dx} \right)^2 \right)}{\alpha I\,(u_1) + (1-\alpha)\,I\,(u_2)} + \frac{(1-\alpha) \left[1 - \alpha + \frac{\alpha}{\mu} \right] I \left(\left(\frac{du_2}{dx} \right)^2 \right)}{\alpha I(u_1) + (1-\alpha)\,I(u_2)} \equiv I_1 + I_2.$$

$$\tag{9.49}$$

Applying the equality (9.47), we have

$$I_1 = \alpha \frac{I \left(\left(\frac{du_1}{dx} \right)^2 \right)}{I\,(u_1)} = \alpha T(u_1), \tag{9.50}$$

$$I_2 = (1-\alpha) \frac{I \left(\left(\frac{du_2}{dx} \right)^2 \right)}{I\,(u_2)} = (1-\alpha)\,T(u_2) \tag{9.51}$$

and consequently the inequality (9.46) holds.

To conclude this section we note that a new problem of optimal design of thin-walled axisymmetric shells, loaded by internal pressure, has been formulated and investigated taking into account strength constraints of modern fracture mechanics. The mass effectiveness of the shell was taken as a maximized functional and the shape of the shell (meridian equation) and the thickness distribution along the meridian were design variables. Joint applications of two control functions (design variables) allowed us not only to obtain a significant increase of mass effectiveness of the shell, but also to simplify the analysis of the optimization problem.

Then it was shown that the optimal solution of the design problem depends on one dimensionless parameter characterizing the degree of lengthening or geometric compression of the axisymmetric shell. Finding the optimal shape and the optimal thickness distribution of an axisymmetric shell was reduced to two closed-form solutions: the first characterized by the critical value of normal stress in the meridian direction and the second by the critical stress in the circumferential direction. We found the range of the problem parameter for which the solution of the first type is optimal. It was also shown that the corresponding problem is convex and consequently the solution is unique for the considered interval of the problem parameter. We proved that there is no solution of the second type.

Chapter 10
Gravity Forces and Snow Loading

10.1 Equilibrium of Shells Under Gravity Forces

Consider an axisymmetric shell with length L, thickness $h = h(x)$ and radius $r = r(x)$ where $0 \leq x \leq L$. The shell is loaded by gravity forces in the axial direction (see Fig. 10.1).

The membrane stresses are

$$\sigma_\varphi = \frac{N_\varphi}{h}, \quad \sigma_\theta = \frac{N_\theta}{h}, \quad \sigma_{\varphi\theta} = \frac{N_{\varphi\theta}}{h} = 0. \tag{10.1}$$

Membrane forces N_φ and N_θ are found using the equilibrium equation along normal to the middle surface

$$\frac{N_\varphi}{r_\varphi} + \frac{N_\theta}{r_\theta} = gh \, \cos\varphi, \tag{10.2}$$

$$\cos\varphi = \frac{\frac{dr}{dx}}{\sqrt{1 + \left(\frac{dr}{dx}\right)^2}} \tag{10.3}$$

and the equation of equilibrium in the axial direction (along the x-axis)

$$R + 2\pi g \int_0^x rh\sqrt{1 + \left(\frac{dr}{dx}\right)^2}\, dx = \frac{2\pi r N_\varphi}{\sqrt{1 + \left(\frac{dr}{dx}\right)^2}}, \tag{10.4}$$

where g is specific weight of the shell material. The reaction force R, applied to the shell at the point $x = 0$, is

$$R = -2\pi g \int_0^L rh\sqrt{1 + \left(\frac{dr}{dx}\right)^2}\, dx. \tag{10.5}$$

N.V. Banichuk and P.J. Neittaanmäki, *Structural Optimization with Uncertainties*, Solid Mechanics and Its Applications 162, DOI 10.1007/978-90-481-2518-0_1,
© Springer Science+Business Media B.V. 2010

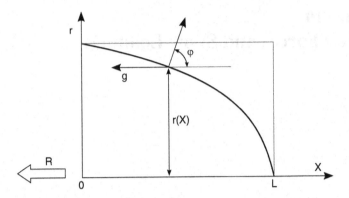

Fig. 10.1 Meridian plane of axisymmetric shell loaded by gravity forces

The relations (10.2)–(10.5) give

$$N_\varphi(x) = -\frac{g}{r}\sqrt{1 + \left(\frac{dr}{dx}\right)^2} \int_x^L rh\sqrt{1 + \left(\frac{dr}{dx}\right)^2}\, dx, \qquad (10.6)$$

$$N_\theta = gh\frac{dr}{dx} - \frac{g\left(\frac{d^2r}{dx^2}\right)\int_x^L rh\sqrt{1 + \left(\frac{dr}{dx}\right)^2}\, dx}{\sqrt{1 + \left(\frac{dr}{dx}\right)^2}}. \qquad (10.7)$$

The expression (10.6) shows that the normal force $N_\varphi(x)$ can take only negative values. This means that the normal stresses in the meridian direction $\sigma_\varphi(x)$ are compressive. But the normal membrane forces $N_\theta(x)$, as is seen from (10.7), can change its sign. Here can be compressive and tensile stresses.

10.2 Weight Minimization

Consider the following optimization problem. Find design variables $r = r(x)$ and $h = h(x)$ such that the weight functional attains its minimum

$$J = 2\pi g \int_0^L rh\sqrt{1 + \left(\frac{dr}{dx}\right)^2}\, dx \to \min \qquad (10.8)$$

under strength conditions

$$-\sigma_1 \le \frac{N_\theta}{h} \le \sigma_2, \qquad (10.9)$$

$$-\sigma_1 \leq \frac{N_\varphi}{h} \qquad (10.10)$$

and geometric constraints, imposed on the shell sizes, thickness distribution and on the contained volume

$$r(0) = r_1, \quad r(L) = 0, \qquad (10.11)$$

$$r_g(x) \leq r(x) \leq r_{max}, \qquad (10.12)$$

$$h_{min} \leq h(x) \leq h_{max}, \qquad (10.13)$$

$$V = \pi \int_0^L r^2 dx \geq V_0, \qquad (10.14)$$

where $r_g(x)$ is a given function and $\sigma_1, \sigma_2, r_{max}, h_{min}, h_{max}, r_1, V_0$ are given positive constants (problem parameters). For convenience of computations we transform basic relations of the optimization problem (10.8)–(10.14) to dimensionless form. We introduce the following dimensionless variables and parameters:

$$\tilde{x} = \frac{x}{L}, \quad \tilde{r} = \frac{r}{r_{max}}, \quad \tilde{h} = \frac{h}{h_{max}},$$

$$\tilde{h}_{min} = \frac{h_{min}}{h_{max}}, \quad \tilde{r}_g = \frac{r_g}{r_{max}}, \quad \tilde{J} = \frac{J}{2\pi g h_{max} r_{max} L},$$

$$\tilde{V} = \frac{V}{r_{max}^2 L}, \quad \tilde{V}_0 = \frac{V_0}{r_{max}^2 L}, \quad \tilde{N}_\varphi = \frac{N_\varphi}{gLh_{max}}, \qquad (10.15)$$

$$\tilde{N}_\theta = \frac{N_\theta L}{gh_{max} r_{max}^2}, \quad \tilde{\sigma}_1 = \frac{\sigma_1}{\sigma^*}, \quad \tilde{\sigma}_2 = \frac{\sigma_2}{\sigma^*},$$

$$\alpha = \frac{r_{max}^2}{L^2}, \quad \beta = \frac{g r_{max}^2}{L\sigma^*}, \quad \gamma = \frac{gL}{\sigma^*},$$

where σ^* is a given positive value. In dimensionless variables, the minimized functional (weight of the shell) and constraints for force and geometric characteristics are

$$J = \int_0^1 rh\sqrt{1 + \alpha \left(\frac{dr}{dx}\right)^2}\, dx, \qquad (10.16)$$

$$\psi_1 = \frac{\gamma}{hr}\sqrt{1 + \alpha \left(\frac{dr}{dx}\right)^2} \int_x^1 rh\sqrt{1 + \alpha \left(\frac{dr}{dx}\right)^2}\, dx - \sigma_1 \leq 0, \quad (10.17)$$

$$\psi_2 = \frac{\beta}{h}\left(rh\frac{dr}{dx} - \frac{\frac{d^2r}{dx^2}\int_x^1 rh\sqrt{1+\alpha\left(\frac{dr}{dx}\right)^2}dx}{\sqrt{1+\alpha\left(\frac{dr}{dx}\right)^2}}\right) - \sigma_2 \le 0, \tag{10.18}$$

$$\psi_3 = -\frac{\beta}{h}\left(rh\frac{dr}{dx} - \frac{\frac{d^2r}{dx^2}\int_x^1 rh\sqrt{1+\alpha\left(\frac{dr}{dx}\right)^2}dx}{\sqrt{1+\alpha\left(\frac{dr}{dx}\right)^2}}\right) - \sigma_1 \le 0, \tag{10.19}$$

$$r_g \le r(x) \le 1, \quad 0 \le x \le 1, \tag{10.20}$$

$$h_{\min} \le h(x) \le 1, \quad 0 \le x \le 1, \tag{10.21}$$

$$\psi_4 = V_0 - \pi\int_0^1 r^2 dx \le 0. \tag{10.22}$$

where we have omitted the tilda throughout.

To solve the optimization problem (10.16)–(10.22) for the axisymmetric shell under the action of force of gravity and the reaction force, we shall use the method of penalty functions in combination with a genetic algorithm. For this purpose we introduce the augmented functionals

$$J^a = J + \sum_{i=1}^4 \mu_i J_i, \tag{10.23}$$

where μ_i are arbitrary positive constants and J_i are penalty functions defined as

$$J_i = \int_0^1 \Psi_i dx, \quad i = 1, 2, 3, \tag{10.24}$$

$$\Psi_i = \begin{cases} 0, & \text{if } \psi_i \le 0, \\ \psi_i, & \text{if } \psi_i > 0, \end{cases} \tag{10.25}$$

$$J_4 = \begin{cases} 0, & \text{if } \psi_4 \le 0, \\ \psi_4, & \text{if } \psi_4 > 0. \end{cases} \tag{10.26}$$

Thus the original problem of optimal design is reduced to finding of design variables $r = r(x), h = h(x)$ satisfying the inequalities (10.20), (10.21) and minimizing the augmented functional (10.23), where the functional J and J_i ($i = 1, 2, 3, 4$) are given by (10.16) and (10.24)–(10.26).

10.3 Shape and Thickness Optimization by Genetic Algorithm

To minimize the augmented functional (10.24)–(10.26) under the constraints (10.20) and (10.21), we apply the evolutionary method of optimization based on a genetic algorithm, which proves to be efficient in searching for the global minimum. We use the conventional terminology [Gol89, HJK$^+$00, HM03, MMM99, MMNP99, PJPO01, BIMS05a, BIMS05b, Gol89, QPPW98, HG92] to describe the computational procedure of this method.

We consider a population changing in the process of iteration and consisting of several individuals representing admissible shell projects. We assume that the number of individuals n_{ind} in each population does not vary in the process of iterations. For convenience, we let the number n_{ind} be even. We also assume that the range $[0, 1]$ is divided into equal parts by points x_i ($i = 1, 2, \ldots, n_{div}$). Each jth individual of the population (the jth admissible project of the shell to be optimized) is described by two sets of point nodes $r(j, i)$ and $h(j, i)$ representing the values of the design variables $r(x)$ and $h(x)$ at the point with number i. Using the algorithm described below, we seek the "best" individual, i.e. two set of variables $r(j, i)$ and $h(j, i)$ for which the functional (10.23)–(10.26) attains its global minimum under the constraints (10.20) and (10.21).

The first step of the algorithm consists in the initialization of the population, which means that $r(j, i)$ and $h(j, i)$ are assigned a random value from the intervals $\left[r_g(i), 1 \right]$ and $[h_{min}, 1]$. The elements $r(j, i)$ and $r(j, n_{div})$ are assigned some fixed values, given by the boundary conditions. Then we calculate the values $J^a(j)$ and find the individual with the least value of this functional.

At the second step of the algorithm, we choose $n_{ind}/2$ pairs of individuals (parents). The first parent ("a") is chosen as follows. We introduce a certain number n^T, choose n^T individuals randomly, and keep that individual with the minimal value of J^a. Similarly, we choose the second parent ("b") of the first pair. As a result we will have $n_{ind}/2$ of such pairs.

At the third step of the algorithm we obtain n_{ind} individuals-offsprings, which forms a new population. To this end, we introduce a certain constant number p_{co} (probability of crossing) such that $0 < p_{co} < 1$, and generate a random number p_r from the interval $[0, 1]$ and a random integer m ($m \in [1, 2, \ldots, n_{div}]$) for each parental pair. If $p_r \geq p_{co}$ then the values of design variables of the offsprings at the nodes $i = 1, \ldots, m$ are copied from their parents "a" and "b", and the values of these variables at the nodes $i = m + 1, \ldots, n_{div}$ are "crossed". The latter means that the values of the offspring "a" are assigned corresponding values of the parent "b", and vice versa.

The fourth step of the algorithm is the mutation of the new population. This step is necessary to avoid being trapped at a local extremum of the functional. To realize the mutation procedure, we introduce a sufficiently small ($\sim$0.05) parameter p_m (mutation probability). Then, for each node of each individual in the population, we generate a random number p_r in the interval $[0, 1]$ and if $p_r \leq p_m$ then the value of the design variable at this particular node is replaced by an arbitrary value satisfying the given restriction. At the nodes with numbers $i = 1$ and $i = n$, the mutation

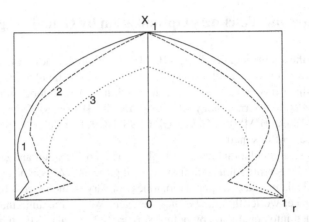

Fig. 10.2 Optimal shape of a dome-like shells of revolution

procedure is not performed for the variables $r(j, i)$. For the new population thus obtained, we calculate the functionals $J^a(j)$ and choose the best individual; then we pass to the second step of the algorithm. We note that if the best offspring in the new population turns out to be "worse" than the best parent from the preceding population, then it is replaced with this parent. Thus, the search for the global minimum is monotonic.

This algorithm was used to obtain optimal distributions of shapes and thickness of dome-like shells for the following parameters values: $n_{ind} = 10$, $n_{div} = 11$, $n^T = 4$, $p_{co} = 0.5$ and $p_m = 0.05$. The program stopped working after 10,000 populations had been generated. The computations were performed for $\alpha = 1$, $\beta = 0.1$, $\gamma = 0.1$, $h_{min} = 0.1$, $r_g(x) = 0$, $\sigma_1 = \sigma_2 = 1$ and $\mu_i = 1$ $(i = 1, \ldots, 4)$.

In Fig. 10.2, curves 1, 2 and 3 show the optimal shapes of the shells for the boundary conditions

$$r(0) = 1, \quad r(1) = 0 \tag{10.27}$$

and for the volumes enclosed by the shell

$$V_0 = 2, \quad V_0 = 1.5 \text{ and } V_0 = 1, \tag{10.28}$$

respectively. The dependence of the value of the functional J^a on the population number n_{pop}, which characterizes the convergence of the computation process, is shown in Fig. 10.3 by the corresponding curves with numbers 1, 2 and 3.

The optimal shapes of shells for $V_0 = 2$ and two versions of the boundary condition

$$r(0) = 1 \text{ and } r(0) = 0.9$$

are shown in Fig. 10.4 (curves 1 and 2).

In Fig. 10.5, curves 1 and 2 correspond to the optimal shapes of shells for

$$r(0) = 0.8, \quad r(0) = 0.5, \quad (V_0 = 2)$$

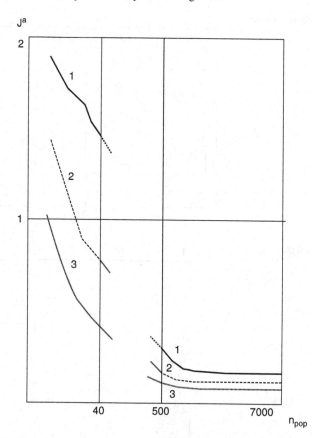

Fig. 10.3 Dependence of the functional J^a on the number of population n_{pop}

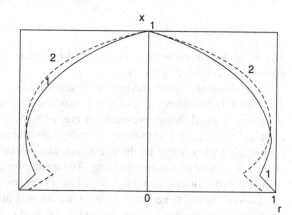

Fig. 10.4 Optimal shell shapes

Fig. 10.5 Optimal dome-like shape

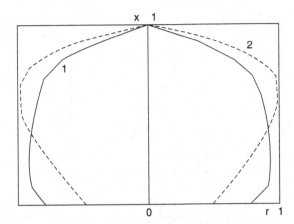

Fig. 10.6 Optimal thickness distributions

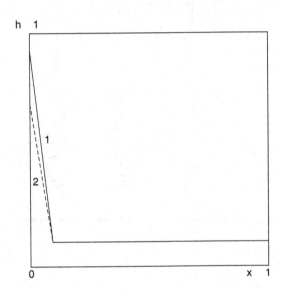

The computations show that, for values of $r(0)$ close to 1, the shape of the optimal dome is not convex (Fig. 10.4).

In all these examples, the optimal distributions of shell thicknesses $h(x)$ were also found. In Fig. 10.6, curves 1 and 2 show the thickness distributions corresponding to the optimal shapes presented in Fig. 10.5. We note that the thickness of the optimal shell almost everywhere coincides with the constraint h_{min} except for the supporting region where its thickness increases noticeably; this can be interpreted as the appearance of a supporting ring. The qualitative characteristics of the optimal thickness distributions are typical for all the examples considered.

In several cases, the computations of the optimal dome shape and thickness distribution also took into account the additional load consisting in an increase in the weight of slanted parts; this imitates snow adhesion.

Fig. 10.7 Optimal dome-like shapes of shells computed for different loading cases

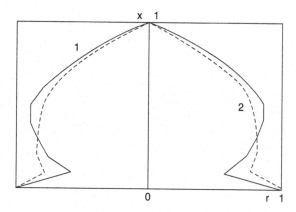

Fig. 10.8 Optimal thickness distributions

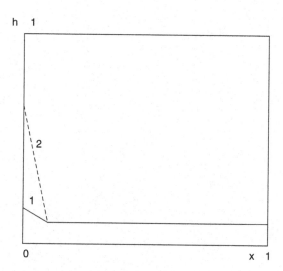

In realizing the genetic algorithm and calculating the weight functional, the parts of the shell satisfying the condition

$$dr/dx < -1$$

were taken to be 1.5 times heavier. Fig. 10.7 compares the shell optimal shapes obtained without the snow load (curve 1) and with the additional load (curve 2) for the values $V_0 = 1.5$ and $r(0) = 1$. One can see that the length of slanted parts of curve 2 is shorter than those of curve 1. The optimal thickness distribution $h(x)$ for the dome shapes presented in Fig. 10.7 is illustrated in Fig. 10.8 by curves 1 and 2. The snow load leads to thickening of the "supporting" ring.

Chapter 11
Damage Characteristics and Longevity Constraints

11.1 Transformations of the Longevity Constraint

Some aspects of optimal design of structures under cyclic loading have been discussed in Chapter 5 taking into account crack appearing and growth. The optimization problems contained a constraint, the number of cyclic before fracture; we call this the longevity constraint. In this section and the next we present some results of optimization of beams, plates, shells and beam-like structures [Ban97, Ban98, BN07, BN08a, BN08b].

It is assumed that the external loads are quasi-statically cyclic within given limits proportionally to the loading parameter p, i.e.

$$0 \leq p_{min} \leq p \leq p_{max},$$

where p_{min} and p_{max} are given positive values. The following damage scenario will be taken into account: an opening crack arises in the structure during its manufacturing or exploitation; the material of the structure is quasi-brittle; the crack is rectilinear; and its initial length l_i is small. Note that the initial length and location of the crack are unknown beforehand. After application of cyclic loading, the initial crack goes ahead and its length l monotonically increases as a function of the number of cycles n. The process of the fatigue crack growth under cycling loading is described by Paris' law

$$\frac{dl}{dn} = C(\Delta K_1)^m, \tag{11.1}$$

$$l_i \leq l \leq l_{cr}, \ 0 \leq n \leq n_{cr} \tag{11.2}$$

and is continued up to the moment when n, the number of cycles, and l, the length of the crack, attain their critical values:

$$n = n_{cr}, \ l = l_{cr}. \tag{11.3}$$

N.V. Banichuk and P.J. Neittaanmäki, *Structural Optimization with Uncertainties*, Solid Mechanics and Its Applications 162, DOI 10.1007/978-90-481-2518-0_1,
© Springer Science+Business Media B.V. 2010

After that, there is unstable crack growth (catastrophic fracture of the structure). Here C and m ($2 \leq m \leq 4$) are some material constants, K_1 is the stress intensity factor for opening cracks and the increment ΔK_1 is given by

$$\Delta K_1 = (K_1)_{\max} - (K_1)_{\min} = (K_1)_{p=p_{\max}} - (K_1)_{p=p_{\min}}. \qquad (11.4)$$

Paris' law and the Griffith criteria [Che79, Hel84, Hut79, GE94]

$$K_1 = K_{1C}$$

are used to describe the crack growth and to evaluate the critical values. The safety condition, well known from quasi-brittle fracture mechanics, is

$$K_1 < K_{1C} \qquad (11.5)$$

and is approximated by the modified inequality constraint [Bol61, Bol69, Str47]

$$K_1 \leq K_{1\varepsilon}, \quad K_{1\varepsilon} = K_{1C} - \varepsilon, \quad \varepsilon \geq 0. \qquad (11.6)$$

Here we introduce a small positive parameter ε and a new term $K_{1\varepsilon}$ with limit K_1. This approximation is convenient in the optimization procedures.

Note that the expression

$$K_1 = \begin{cases} \kappa p \sigma_n \sqrt{\pi l}, & \sigma_n > 0, \\ 0, & \sigma_n \leq 0, \end{cases} \qquad (11.7)$$

for the stress intensity factor will be used for the crack when it is small enough and its length l satisfies the inequalities

$$l_i \leq l \leq l_{cr} \ll r_m. \qquad (11.8)$$

Here r_m is a characteristic size, and the coefficient κ depends on the type of crack (internal crack, surface crack, through the thickness crack). The value $p\sigma_n$ in (11.7) is the normal stress in the uncracked structure at the place of crack location. Subscript n means that the stress acts in the direction normal to the crack banks.

The fracture criterion

$$(K_1)_{l=l_{cr}, p=p_{\max}} = K_{1\varepsilon} \qquad (11.9)$$

and the formula (11.7) imply

$$l_{cr} = \frac{1}{\pi} \left(\frac{K_{1C}}{\kappa p_{\max} \sigma_n} \right)^2. \qquad (11.10)$$

To find the expression for the critical number of cycles, we integrate (11.1), (11.2) and determine l_{cr} from the relation (11.10). Performing some elementary transformations we will have

$$n_{cr} = \frac{a_1}{l_i^{\frac{m}{2}-1} \sigma_n^m} \left[1 - \left(a_2 \sigma_n^2\right)^{\frac{m}{2}-1}\right],$$

$$a_1 = \frac{1}{\left(\frac{m}{2} - 1\right) C \kappa^m \pi^{\frac{m}{2}} (\Delta p)^m}, \tag{11.11}$$

$$a_2 = \frac{l_i \pi \kappa^2 p_{max}}{K_{1C}^2}, \quad \Delta p = p_{max} - p_{min}.$$

As it can be seen from (11.11) the value n_{cr} is essentially dependent on σ_n, and, consequently, the dependence of the critical number of cycles n_{cr} on geometrical design parameters that define the shape of the body is realized by means of the dependence of the normal stresses on these parameters. We will consider the ranges of mechanical and geometric parameters in (11.11) for which $n_{cr} > 0$. For this case the dependences of n_{cr} on l_i and σ_n, as can be seen from (11.11), are monotonic. The value n_{cr} is decreasing when we increase the values l_i and σ_n. Thus the minimum of critical number of cycles is attained for

$$\max_{\omega} l_i = l_m, \tag{11.12}$$

$$\max_{\omega} \sigma_n = \sigma_{nm} \tag{11.13}$$

and equals to

$$\min_{\omega} n_{cr} = \frac{a_1}{l_m^{\frac{m}{2}-1} \sigma_{nm}} \left[1 - \left(a_2 \sigma_{nm}\right)^{\frac{m}{2}-1}\right]. \tag{11.14}$$

For surface cracks, $\kappa = 1.12$, and the vector ω characterizing the crack has two components: length of the crack and coordinate of the point on the body surface. Here we suppose that the body is two-dimensional and the surface crack is oriented normally to the surface. The value σ_{nm} must be less than or equal to its limit corresponding to the given critical number of cycles n_0. This limit is derived from the equation

$$\min_{\omega} n_{cr} = n_0 \tag{11.15}$$

and is denoted by σ_0. The longevity constraint (5.44) is

$$\max_{\omega} \sigma_n \leq \sigma_0, \tag{11.16}$$

$$\sigma_0 \equiv \chi(\kappa, m, C, K_{1C}, l_m, \Delta p, p_{max}, n_0),$$

where χ is a function of material constants m, C, K_{1C}, admissible minimum number of cycles n_0 and maximum initial crack length l_m and other problem parameters characterized by the loading process. Consider the case when the process of fatigue crack growth under cycling loading is described Paris' law ($m = 4$) in (11.1) and (11.11). For $m = 4$ the expression (11.11) for critical number of cycles n_{cr} is

$$n_{cr} = \frac{1}{\pi^2 l_i C \left(\kappa \sigma_n \Delta p\right)^4} \left(1 - \pi l_i \left(\frac{\kappa p_{max} \sigma_n}{K_{1C}}\right)^2\right)$$ (11.17)

and the function χ, used in the longevity condition (11.16), is given by the explicit formulas

$$\chi = \left(b_1 \left(-1 + \sqrt{1 + b_2}\right)\right)^{1/2},$$

$$b_1 = \frac{P_{max}^2}{2\pi \kappa^2 C \left(\Delta p\right)^4 K_{1C}^2 n_0},$$ (11.18)

$$b_2 = \frac{4 n_0 C K_{1C}^4 \left(\Delta p\right)^4}{l_m P_{max}^4}.$$

11.2   Investigation of Optimal Design

First consider the problem of minimum mass design of the beam under a longevity constraint. The beam has length L, ($0 \leq x \leq L$) and rectangular cross-section with thickness h and width b. Applied loads are proportional to the parameter p. Cracks may appear at the upper and lower surfaces ($\zeta = \pm h/2$) of the beam. Normal stresses acting on the tensile surface of the uncracked beam (for definiteness $\zeta = h/2$) are described by the formulas

$$\sigma_x = \left(\frac{|M|\zeta}{I}\right)_{\zeta = \frac{h}{2}} = \frac{6|M|}{bh^2},$$ (11.19)

$$M = pM_0(x), \quad 0 \leq x \leq L$$ (11.20)

and, consequently, the following relation for σ_n must hold:

$$\sigma_n = \left(\frac{6|M|}{bh^2}\right)_{p=1} = \frac{6|M_0|}{bh^2},$$ (11.21)

where M is the bending moment acting at the considered cross-section of the beam. In general case the value σ_x is an implicit function of the design parameters. For a statically determinate beam or beam-like structures, the bending moment M does not depend on the material constants or the geometric parameters of the cross-section, i.e. $M = pM_0(x)$, where the distribution $M_0(x)$ is considered as a given function. Thus the normal tensile stresses at the beam surface are given by an explicit function of geometric parameters. If we assume that the width b is a given constant and the thickness $h = h(x)$ is the unknown design variable, then the longevity constraint (11.16) takes the form

$$\max_x \sigma_n = \frac{6}{b} \max_x \left(\frac{|M_0(x)|}{h^2(x)} \right) \le \sigma_0. \tag{11.22}$$

The minimum of the integral functional (mass of the beam)

$$J = \rho b \int_0^L h(x) dx \to \min_h \tag{11.23}$$

under the constraint (11.22), is realized for the following optimum distribution of structural thickness:

$$h_*(x) = (6|M_0(x)|/b\sigma_0)^{1/2}. \tag{11.24}$$

Here ρ is the density of the material. If the bending moment $M(x)$ tends to zero $(M_0(x) \to 0)$ when x tends to some point x_0 $(x_0 \in [0, L])$, then the structural thickness, as can be seen from (11.24), also tends to zero, and the geometric constraint (11.8) $(l_i \le l_m \le l_{cr} \ll \min h(x)$ for our case) is violated. In this case the additional geometric constraint

$$h(x) \ge h_{\min} \gg l_{cr} \tag{11.25}$$

is required. The corresponding optimum thickness distribution takes the form

$$h_* = h_*(x) = \max \left\{ h_{\min}, \sqrt{\frac{6|M_0(x)|}{b\sigma_0}} \right\}, \tag{11.26}$$

where the max-operation in (11.26) means finding the maximum for two values written in braces for each $x \in [0, L]$.

Consider the problem arising when we wish to determine the optimum shape of the hole in an infinite plate, assuming that the boundary of the hole Γ has no external forces applied to it and the plate is subjected to cyclic loads applied at infinity [BMN00, Mus53], i.e.

$$(\sigma_s)_\Gamma = 0, \quad (\sigma_{st})_\Gamma = 0,$$
$$(\sigma_x)_\infty = p\sigma_1^\infty, \quad (\sigma_y)_\infty = p\sigma_2^\infty, \tag{11.27}$$
$$(\sigma_{xy})_\infty = 0, \quad 0 \le p_{\min} \le p \le p_{\max},$$

where $\sigma_1^\infty, \sigma_2^\infty, p_{\min}, p_{\max}$ are given positive constants and s, t denote, respectively, the directions normal and tangent to the curve Γ. The optimization problem consists in finding the optimal shape of the contour Γ such that the minimum of the critical number of cycles n_{cr} is maximized

$$J = \max_{\Gamma} \min_{Q \in \Gamma} \min_{l \leq l_m} n_{cr}, \qquad (11.28)$$

where Q is a point of the boundary Γ (origin of the surface crack).

Using the same arguments as before, we transform the problem (11.28) to minimization of the maximal tensile stress at the contour Γ

$$\min_{\Gamma} \max \sigma_t. \qquad (11.29)$$

Note that for the surface crack, the normal to its bank coincides with the direction tangential to the contour Γ and $\sigma_n = \sigma_t$. Thus the optimization problem for a plate with a hole, and surface cracks arising at its boundary and oriented in the normal to the boundary direction, is reduced to the problem of tension stress minimization, for which the global optimal solution is given by the one-parameter family of holes $\left(x/\sigma_1^{\infty}\right)^2 + \left(y/\sigma_2^{\infty}\right)^2 = \lambda^2$ (λ is an arbitrary constant) with equally-stressed elliptic contours [Ban83, Ban90].

11.3   Loader Crane Optimization with Longevity Constraints

Consider the problem of minimum mass design for the loader crane structure shown in Fig. 11.1. The structure consists of substructure 1 – column, substructure 2 – first boom (lifting boom), substructure 3 – second boom (shifting boom) and substructure 4 – boom extension (sliding boom) and appendages (hydraulic system, electric system, interconnections).

All the elements of the loader crane are placed in the xy-plane of an orthogonal coordinate system. (The x-axis is oriented along the substructures 2–4, the y-axis is in the orthogonal direction and connected at the joints.) We use the following notation for the masses of separate parts of the loader crane structure: J_c – mass of

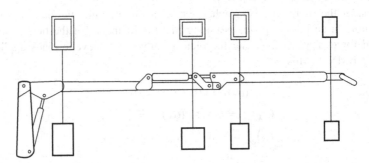

Fig. 11.1 Optimization of the loader crane substructures 2–4 (lifting boom, shifting boom, sliding boom)

the column, J_{b1} – mass of the first boom, J_{b2} – mass of the second boom, J_{be} – mass of the boom extension, J_a – mass of the appendage and J – total mass:

$$J = J_c + J_{b1} + J_{b2} + J_{be} + J_a. \tag{11.30}$$

The structure is loaded by a cyclic force acting in the y-direction at the free end of the boom extension. All the substructures have hollow cross-sections (see Fig. 11.1) and in bending the maximum tensile normal stresses are attained at the surface of the beam-like structural elements. The tensile stress σ_t is evaluated by the following formula:

$$\sigma_t = \frac{|M| H}{2I}, \quad I = \frac{hH^2(H + 3b)}{b}. \tag{11.31}$$

Here we investigate crack appearance at the external surface of the beam-like sub-structures 1–4, and use the notations H, b and h, respectively, for height, width and thickness of the wall, and M for external bending moment at the rectangular cross-section. Variables b and are taken as constant along each of the substructures 1–4 (piece-wise constant along the built-up structure of the loader crane). Variable H is linearly varied along the substructures 1, 2 and is taken as a constant along the substructures 3, 4.

11.3.1 First Example

In the first example the values H and h are taken as unknown design variables for substructures 2–4 and found under constraint $h \geq h_{min}$ ($h_{min} = 0.004m$). The geometric parameters of the column are known and not varied in the design process. This means that we vary only the design parameters of the first boom, second boom and boom extension. The optimization problem consists of minimization of the mass of the system of substructures 2–4.

$$J = J_{b1} + J_{b2} + J_{be} \to \min_{H,h} \tag{11.32}$$

under longevity constraint and additional rigidity constraint (elastic deflection of the point of load application is less than or equal to a given value: $w \leq w_* = 0.074\,m$).

For the loader crane material (steel) the material constants [Hel84] are

$$K_{1C} = 27MNm^{3/2}, \quad m = 3.1,$$
$$C = 3.2 \cdot 10^{-12} MN^{-m} m^{(3m+2)/2}.$$

Other important parameters were taken as

$$l_i = l_m = 10^{-5}m, \quad p_{min} = 0,$$
$$n_0 = 1.2475 \cdot 10^7, \quad \kappa = 1.12.$$

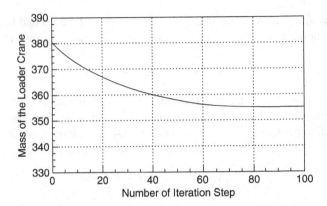

Fig. 11.2 Total mass as a function of the number of iteration steps

The loader crane is statically indeterminate, and the values of the bending moment M used in (11.31) depend on geometric and material parameters. The analysis was performed numerically by using finite-element approximations (see, for example, [HN88, NST06, NR04]). The optimal design was found by successive optimization algorithm [Ban90] based on the gradient projection method. The optimal design was found after 100 steps, and a meaningful reduction in mass was attained (Fig. 11.2).

Figure 11.1 shows results of optimization of the geometric parameters. The greatest and least cross-sections correspond respectively to the initial and optimal designs. Figure 11.1 shows that the cross-sections of the substructures increase, while the walls have become thinner, and attain the minimum admissible value ($h_{min} = 0.004 m$). The relative loss of mass of the complete crane was $\eta = 7.1\%$. The relative loss of mass for the optimized part was

$$\eta_J = \frac{J_{init} - J_{opt}}{J_{init}} \times 100\% = \frac{\Delta J_{b1} + \Delta J_{b2} + \Delta J_{be}}{(J_{b1} + J_{b2} + J_{be})_{init}} \times 100\% = 17.6\%.$$

11.3.2 Second Example

Consider another case when J_{b1}, J_{b2}, J_{be} are constant and J_C is varied

$$J = J_c \to \min_{H,h}. \qquad (11.33)$$

This means that we vary only the column design parameters. The cross-sectional geometric parameters H, h ($h \geq h_{min} = 0.004\,m$) are taken as design variables, while the longevity and rigidity ($w \leq w_* = 0.074\,m$) requirements are the same as before. The optimal design was found after ten steps, and the reduction in total mass J is shown in Fig. 11.3 as a function of iteration step number.

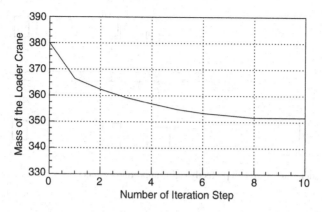

Fig. 11.3 Decreasing of the total mass of the loader crane in the case of optimization of substructure 1 (*column*)

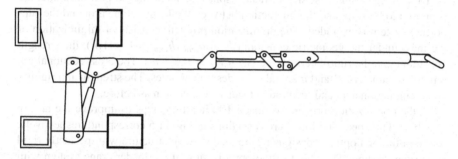

Fig. 11.4 Loader crane structure. Optimization of substructure 1 (*column*)

The results of optimization of the geometric parameters H, h for substructure 1 (column) are presented in Fig. 11.4. The relative loss of mass referred to the total mass was $\eta = 7.9\%$, while the relative loss of mass for the substructure 1 was

$$\eta_J = \frac{J_{init} - J_{opt}}{J_{init}} \times 100\% = \frac{\Delta J_c}{(J_c)_{init}} \times 100\% = 56.5\%.$$

This is a significant mass reduction.

11.3.3 Third Example

Consider the case when

$$J = J_{b1} + J_{b2} + J_{be} \to \min_{H,h} \tag{11.34}$$

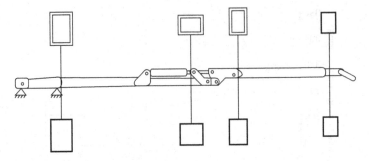

Fig. 11.5 Loader crane arm with perfectly rigid supporting system

and the column and its supporting system are perfectly rigid. In this case the loader crane system can be modeled as the beam-like structure shown in Fig. 11.5.

To investigate this case the vertical displacements at the supporting points are supposed to be zero, and the interaction between the loader crane arm and the supporting system is excluded. The optimization problem consists of minimization of the mass under the geometric constraint $h \geq h_{\min}$ ($h_{\min} = 0.004\,m$), the strength constraint and the rigidity constraint ($w \leq w_* = 0.074\,m$). The cross-sectional geometric parameters H and h are taken as design variables. The strength and rigidity constraint parameters and the gradient steps were chosen as before.

In the optimization process we made 100 iteration steps to improve the design variables. The upper and lower cross-sections in Fig. 11.5 correspond, respectively, to the initial and optimal designs of the loader crane structure with perfectly rigid supporting system. The effect of the optimization of the loader crane system with respect to the initial mass of the designed substructures was evaluated as

$$\eta_J = \frac{\Delta J_{b1} + \Delta J_{b2} + \Delta J_{be}}{(J_{b1} + J_{b2} + J_{be})_{init}} \times 100\% = 19.1\%$$

The methodology developed in this section for problem formulation and solution is general enough to be used for optimization of various brittle and quasi-brittle bodies, structural elements and built up structures. All the investigations have been performed in the framework of the minimax or guaranteed approach, taking into account various initial crack sizes and locations. Analytical results of optimization were presented for beam and plate with hole. Corresponding numerical results of optimization were presented for beam-like built up structures. These results illustrate the effect of optimization.

Chapter 12
Optimization of Shells Under Cyclic Crack Growth

In the previous chapter we considered the problems of optimal design of bodies with surface cracks. In this chapter we present some results of optimization [BIMS05a, BIMS05b, BRS06] of axisymmetric shells containing through the thickness cracks and loaded by various cyclic loads.

12.1 Basic Relations and Optimization Modeling

Consider an elastic shell that has the form of a surface of revolution. The position of the meridian plane is defined by the angle θ, measured from some datum meridian plane, and position of a parallel circle is defined by the angle φ, between the normal to the surface and the axis of rotation (see Fig. 12.1), or by the coordinate x, measured along the axis of rotation; $0 \leq x \leq L$, L is a given value.

An axially symmetrical shape of a middle surface is characterized by a distance $r(x)$ from the axis of rotation to a point of the middle surface (a profile of each cross section of a shell is a circle). This variable $r = r(x)$ and thickness distribution $h = h(x)$ will be considered simultaneously or separately as the design variables. The geometric relations between meridional curvature radius $r_\varphi(x)$, circumferential curvature radius $r_\theta(x)$ and the radius $r(x)$ (see Fig. 12.2) are given by expressions (7.1), (7.2).

We will use also the following geometric relations:

$$r = r_\theta \sin \varphi, \quad \frac{dr}{d\varphi} = r_\varphi \cos \varphi, \quad \frac{dx}{d\varphi} = r_\varphi \sin \varphi. \tag{12.1}$$

The shell is loaded by axisymmetrical forces acting in the meridian planes. The intensities of the external loads, which act in the directions normal and tangential to the meridian, are denoted by q_n and q_φ in Fig. 12.1. The resultant of the total load applied to the parallel circle and acting in the x-direction is denoted by R. Normal membrane forces N_φ, N_θ per unit length and normal membrane stresses σ_φ, σ_θ acting in meridional and circumferential directions are found with the help of (7.6) and the equations [Flu73, SK89]

N.V. Banichuk and P.J. Neittaanmäki, *Structural Optimization with Uncertainties*, Solid Mechanics and Its Applications 162, DOI 10.1007/978-90-481-2518-0_1, © Springer Science+Business Media B.V. 2010

Fig. 12.1 The forces acting on the shell segment

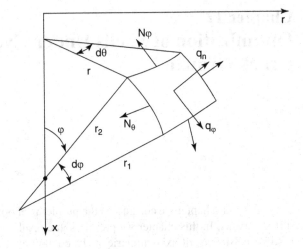

Fig. 12.2 Geometric terminology for the shell

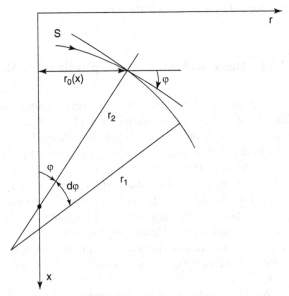

$$\frac{d(rN_\varphi)}{d\varphi} - N_\theta r_\varphi \cos\varphi + rr_\varphi q_\varphi = 0,$$

$$\sigma_\varphi = \frac{N_\varphi}{h}, \quad \sigma_\theta = \frac{N_\theta}{h} \tag{12.2}$$

with corresponding boundary conditions. All external loads q_n, q_φ and their resultants are proportional to the loading parameter p

$$q_n = p\tilde{q}_n, \quad q_\varphi = p\tilde{q}_\varphi, \tag{12.3}$$

$$\tilde{q}_n = (q_n)_{p=1}, \quad \tilde{q}_\varphi = (q_\varphi)_{p=1}. \tag{12.4}$$

It is assumed that the external loads are cyclic and varied quasitatically between given limits, i.e. $0 \leq p_{min} \leq p \leq p_{max}$, where p_{min} and p_{max} are given values. We will take into account that, to compute internal forces and stresses in an uncracked shell, we will use the basic relations of linear shell theory and, consequently, we will have

$$N_j = p\widetilde{N}_j, \quad \sigma_j = p\tilde{\sigma}_j, \tag{12.5}$$

where $j = \varphi, \theta$ and

$$\widetilde{N}_j = (N_j)_{p=1}, \quad \tilde{\sigma}_j = (\sigma_j)_{p=1}. \tag{12.6}$$

It is easily seen that the introduced values $\widetilde{N}_\varphi$, $\widetilde{N}_\theta$, $\tilde{\sigma}_\varphi$ and $\tilde{\sigma}_\theta$ satisfy the same equations as the original values N_φ, N_θ, σ_φ and σ_θ. In what follows, the tilde will be omitted. It is assumed that a through the thickness crack can arise in the shell during its manufacturing or exploitation, and assume that the material of the shell is quasi-brittle. The crack is supposed to be rectilinear, and its length very small with respect to the characteristic geometric sizes of the shell without any restriction on the location of the crack in the shell, its orientation, and its initial length $l_i \leq l_{cr}$. The value l_{cr} determines the moment when a global fracture occurs. In what follows, we will assume that not only the initial crack but also all temporary cracks ($l_i \leq l \leq l_{cr}$) are rectilinear, and that the crack length l is larger than h and is small with respect to the characteristic size r_m of the shell, i.e.

$$h_m \leq l_i \leq l \leq l_{cr} \ll r_m, \tag{12.7}$$

where

$$h_m = \max_{0 \leq x \leq L} h(x),$$

$$r_m = \min \left(\min_{0 \leq x \leq L} r_\varphi(x), \min_{0 \leq x \leq L} r_\theta(x) \right).$$

It is also supposed that the functions $r = r(x)$ and $h = h(x)$ are smooth enough.

The expression

$$K_1 = \begin{cases} p\sigma_n \sqrt{\frac{\pi l}{2}}, & \sigma_n > 0, \\ 0, & \sigma_n \leq 0, \end{cases} \tag{12.8}$$

will be used for the stress intensity factor when the through the thickness crack is small enough and is distant from the shell boundaries. Note that $p\sigma_n$ is the normal stress in the uncracked shell at the crack location. Subscript n means that the stress acts in the direction normal to the crack banks. Using (12.8) and safety condition (11.9), we will have an expression for l_{cr} in the form

$$l_{cr} = \frac{2}{\pi} \left(\frac{K_{1\varepsilon}}{p_{max}\sigma_n} \right)^2. \tag{12.9}$$

It is assumed for subsequent considerations that the possible locations of initial cracks, arising from manufacture and exploitation, are unknown beforehand. In the context of structural design, this leads to essential complications in the computation of n_{cr}, caused by the necessity to analyze a variety of crack locations and orientations, and to solve structural analysis problems. Let us consider only "internal" (not close to the shell boundary) through the thickness cracks, for which it is possible to use estimate (12.8), and characterize the crack by the vector

$$\omega = \{l, x_c, \alpha\}$$

containing the coordinate x_c of the crack center, the length of the crack l, and the angle α setting the crack inclination with respect to the meridian (see Fig. 12.3).

The second coordinate θ_c of the crack midpoint is nonessential and omitted because we consider axisymmetric problems and admit all locations of the crack in parallel direction ($0 \leq \theta_c \leq 2\pi$). If $\alpha = 0$, the crack is oriented in the meridian direction (axial crack) and for $\alpha = \pi/2$ the crack is peripherally oriented in parallel direction.

The following model assumptions will be made:

1. The quasi-brittle shell contains an "internal" through the thickness crack, modeled by a rectilinear notch, and the notch is traction free.
2. The initial length of the crack l_i and its temporary values l up to l_{cr} ($l_i \leq l \leq l_{cr}$) satisfy the condition (12.7).
3. The shell has only one crack, but this crack can be characterized by any vector ω from a given set Λ_ω ($\omega \in \Lambda_\omega$), where Λ_ω is the set of all admissible cracks under initial restrictions.

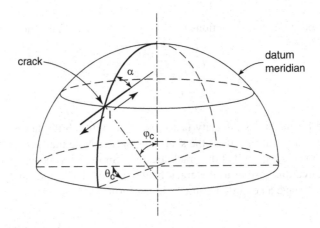

Fig. 12.3 Crack position, orientation and length

Accepted assumptions and available additional data concerning the most dangerous parts of the shell suggest that the set Λ_ω of admissible initial cracks should be considered as given.

Structural longevity is evaluated by the number of load cycles $n = n_{cr}$ for which $l = l_{cr}$ and unstable fracture occurs. In the design process the longevity constraint is taken as $n_{cr} \geq n_0$, where n_0 is a given minimum value. Taking into account incompleteness of the information concerning the possible location, orientation and size of initial cracks, we can rewrite the longevity constraint in the following manner:

$$\min_{\omega \in \Lambda_\omega} n_{cr} \geq n_0. \tag{12.10}$$

For effective analysis of the longevity constraint (12.10) we shall obtain an explicit expression for n_{cr} as a function of the problem parameters. To do this we note that

$$\Delta K_1 = \sigma_n \Delta p \sqrt{\frac{\pi l}{2}}, \quad \Delta p = p_{max} - p_{min} \tag{12.11}$$

and perform integration using (11.1) (11.2) and (12.9), (12.11). We have

$$n_{cr} = n_{cr}\left(l_i, \sigma_n\left(\alpha, x\right)\right) = \frac{\Psi_1\left(l_i, \sigma_n\left(\alpha, x\right)\right)}{\Psi_2\left(l_i, \sigma_n\left(\alpha, x\right)\right)}, \tag{12.12}$$

$$\Psi_1 = \Psi_1\left(l_i, \sigma_n(\alpha, x)\right) = 1 - \left[\frac{\pi l_i}{2}\left(\frac{p_{max}\sigma_n\left(\alpha, x\right)^2}{K_{1\varepsilon}}\right)\right]^{\frac{m}{2}-1}, \tag{12.13}$$

$$\Psi_1 = \Psi_2\left(l_i, \sigma_n\left(\alpha, x\right)\right) =$$

$$= C\left(\frac{m}{2} - 1\right)\left(\frac{\pi}{2}\right)^{\frac{m}{2}} l_i^{\frac{m}{2}-1}\left[p_{max}\sigma_n\left(\alpha, x\right)\left(1 - \frac{p_{min}}{p_{max}}\right)\right]^m. \tag{12.14}$$

It is seen from expressions (12.12)–(12.14) that for any fixed $x \in [0, L]$, the critical number of cycles n_{cr} is a monotonically decreasing function of l_i and σ_n, and, consequently, a minimum of n_{cr} with respect to l_i and σ_n is attained for

$$l = l_{im}, \quad (0 < l_i \leq l_{im} < l_{cr}), \tag{12.15}$$

$$\sigma_n = \max_\alpha \sigma_n(\alpha, x), \tag{12.16}$$

where l_{im} is the given maximum length of considered initial cracks.

As it was shown in Section 7.1 the normal membrane stresses σ_φ and σ_θ are principal and, consequently, we have

$$\sigma_\varphi \leq \sigma_n \leq \sigma_\theta \text{ or } \sigma_\theta \leq \sigma_n \leq \sigma_\varphi.$$

This means that the extremal values σ_n, with respect to the angle of inclination α, are realized for $\alpha = 0$ (meridian direction) and for $\alpha = \pi/2$ (parallel direction).

Thus we find that the minimum of n_{cr} with respect to α is attained when α takes one of two values: $\alpha = 0$ (axial crack) or $\alpha = \pi/2$ (peripheral crack). We have

$$\min_{\omega \in \Lambda_\omega} n_{cr}\left(l_i, \sigma_n\left(\alpha, x\right)\right) = \min_{0 \le x \le L} \hat{n}_{cr}(x), \qquad (12.17)$$

where

$$\hat{n}_{cr}(x) = \min\left\{n_{cr}\left(l_{im}, \sigma_\theta(x)\right), \ n_{cr}\left(l_{im}, \sigma_\varphi(x)\right)\right\}, \qquad (12.18)$$

$$\sigma_\theta(x) = \left(\sigma_n\left(\alpha, x\right)\right)_{\alpha=0}, \quad \sigma_\varphi(x) = \left(\sigma_n(\alpha, x)\right)_{\alpha=\frac{\pi}{2}}. \qquad (12.19)$$

Now the longevity constraint (12.10) can be written in the form

$$\min_{0 \le x \le L} \hat{n}_{cr}(x) \ge n_0 \qquad (12.20)$$

or as a system of two inequalities

$$\begin{cases} n_{cr}(l_{im}, \sigma_\theta(x)) \ge n_0, \\ \qquad\qquad\qquad\qquad\qquad 0 \le x \le L. \\ n_{cr}(l_{im}, \sigma_\varphi(x)) \ge n_0, \end{cases} \qquad (12.21)$$

Application of these constraints can be simplified by means of transformation of (12.21) to obtain direct inequalities imposed on σ_θ and σ_φ. For this purpose we take into account the monotonicity of $n_{cr}(l_{im}, \sigma_n)$ with respect to σ_n, and define the value σ_0 as a root of the algebraic equation

$$n_{cr}\left(l_{im}, \sigma_0\right) = n_0 \qquad (12.22)$$

which is written in explicit form as

$$\frac{1 - \left[\frac{\pi l_{im}}{2}\left(\frac{p_{max}\sigma_0}{K_{1\varepsilon}}\right)^2\right]^{\frac{m}{2}-1}}{C\left(\frac{m}{2}-1\right)l_{im}^{\frac{m}{2}-1}\left(\frac{\pi}{2}\right)^{\frac{m}{2}}\left[p_{max}\sigma_0\left(1 - \frac{p_{min}}{p_{max}}\right)\right]^m} = n_0. \qquad (12.23)$$

For $\sigma_n \le \sigma_0$ we will have

$$n_{cr}\left(l_{im}, \sigma_n\right) \ge n_{cr}\left(l_{im}, \sigma_0\right) = n_0. \qquad (12.24)$$

Consequently, the longevity constraints (12.21) are reduced to the following system of two inequalities

$$\sigma_\varphi \le \sigma_0, \qquad (12.25)$$

$$\sigma_\theta \le \sigma_0. \qquad (12.26)$$

The considered optimization problem consists of finding the shape $r = r(x)$ and the thickness distribution (simultaneously or separately) of the shell, such that the volume of the shell material

$$J = \int_0^{2\pi} \int_{\varphi_0}^{\varphi_f} h r_\varphi r_\theta \sin\varphi d\varphi d\theta = 2\pi \int_{\varphi_0}^{\varphi_1} h r_\varphi r_\theta \sin\varphi d\varphi = 2\pi \int_0^L rh \left(1 + \left(\frac{dr}{dx}\right)^2\right)^{1/2} dx$$

$$(12.27)$$

is minimized, while satisfying the longevity constraint (12.10) or (12.21) or (12.25), (12.26) and additional geometric constraints

$$V = \pi \int_0^L r^2 dx = V_0, \qquad (12.28)$$

$$r(x) \geq r_g(x), \ 0 \leq x \leq L, \qquad (12.29)$$

$$h(x) \geq h_0 \ 0 \leq x \leq L, \qquad (12.30)$$

where V_0, h_0 are given positive constants, $r_g(x) \geq$ is the given function. This restriction does not change the sense of the problem but permits avoidance of possible singularities. The shape of the shell will be called optimal if for any shell with a smaller weight it is possible to select a vector of unknown parameters ω belonging to the admissible set Λ_ω, such that some assigned constraints have been violated.

Note that the case $m = 4$ (typical for metals [Hel84]) Eq. 12.22 is reduced to the form

$$n_{cr} = \frac{1 - \frac{\pi l_{im}}{2} \left(\frac{\sigma_0}{K_{1\varepsilon}}\right)^2}{C \left(\frac{\pi}{2}\right)^2 l_{im} \sigma_0^4 \xi} = n_0, \quad \xi = \left(1 - \frac{p_{min}}{p_{max}}\right)^4 \qquad (12.31)$$

and the expression for the value σ_0 can be found as

$$\sigma_0^2 = b_1 \left(-1 + \sqrt{1 + b_2}\right),$$

$$b_1 = \frac{1}{\pi C \xi K_{1\varepsilon}^2 n_0}, \qquad (12.32)$$

$$b_2 = 4n_0 K_{1\varepsilon}^4 C \frac{1}{l_{im}} \xi.$$

12.2 Optimal Thickness Distribution for Shells of Given Geometry

The considered optimization problem consists of finding the thickness distribution $h = h(x)$, such that the optimized functional (12.27) attains a minimum, while satisfying the geometric constraint (12.30) and the longevity constraint (12.10).

Using the expressions (7.9), (7.10) for the membrane forces N_φ and N_θ, we represent the longevity constraint in the form

$$\max_\varphi \left(\sigma_\varphi(\varphi) = \frac{N_\varphi(\varphi)}{h(\varphi)} \right) \leq \sigma_0,$$

$$\max_\varphi \left(\sigma_\theta(\varphi) = \frac{N_\theta(\varphi)}{h(\varphi)} \right) \leq \sigma_0,$$

$$N_\varphi(\varphi) = -\frac{R(\varphi)}{2\pi r(\varphi) \sin \varphi}, \tag{12.33}$$

$$N_\theta(\varphi) = r_\theta(\varphi) \left(\frac{R(\varphi)}{2\pi r(\varphi) r_\varphi(\varphi) \sin \varphi} + q_n(\varphi) \right).$$

To satisfy the longevity constraints (12.33), it is necessary and sufficient to require that

$$h(\varphi) \geq -\frac{R(\varphi)}{2\pi \sigma_0 r(\varphi) \sin \varphi},$$

$$h(\varphi) \geq \frac{r_\theta(\varphi)}{\sigma_0} \left(\frac{R(\varphi)}{2\pi r(\varphi) r_\varphi(\varphi) \sin \varphi} + q_n(\varphi) \right) \tag{12.34}$$

for $\varphi \in [\varphi_0, \varphi_f]$. Solution of the original optimization problem that was transformed to minimization of the integral (12.27) under the constraints (12.30), (12.34) can be written as

$$h = \max \left\{ h_0, \frac{N_\varphi}{\sigma_0}, \frac{N_\theta}{\sigma_0} \right\} =$$

$$= \max \left\{ h_0, -\frac{R}{2\pi \sigma_0 r \sin \varphi}, \frac{r_\theta}{\sigma_0} \left(\frac{R}{2\pi r r_\varphi \sin \varphi} + q_n \right) \right\}. \tag{12.35}$$

For any fixed $\varphi \in [\varphi_0, \varphi_f]$, operation max in (12.35) means the selection of the large of the three quantities inside the braces.

As an example consider the problem of optimal design of the shell in the form of a torus obtained by rotation of a circle of radius a about a vertical axis. The distance between the centre of the circle and the vertical axis is denoted by b (Fig. 12.4 shows a half of the shell).

The shell is subjected to the cyclic action of uniform internal pressure $q_n = pq^0$ ($q^0 = \text{const}$) that varies between given limits proportionally to parameter p. The forces N_φ, N_θ are obtained by considering the equilibrium of the ring-shaped element of the shell and written as [TW59]

$$N_\varphi = \frac{a}{2} \left(1 + \frac{b}{r} \right) q_n, \quad N_\theta = \frac{a}{2} q_n, \tag{12.36}$$

Fig. 12.4 Optimal design of torus shell

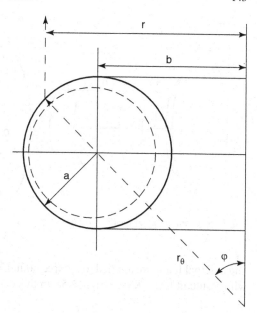

where $r = b + a \sin\varphi$. Taking into account that $N_\varphi > N_\theta$ and using the formula (12.35) we will obtain the optimal thickness distribution in the following form:

$$h = \max\left\{h_0, \frac{N_\varphi}{\sigma_0}\right\} = \max\left\{h_0, \frac{aq^0}{2\sigma_0}\left(1 + \frac{b}{r}\right)\right\}. \qquad (12.37)$$

The optimal thickness distribution is shown in Fig. 12.4 for the torus shell. (12.37) shows that the thickness decreases when the radius r increases.

Note that when $m = 4$ and $n_{cr} \to \infty$ the following representation for σ_0 takes place:

$$\sigma_0 = \left[\sqrt{\frac{\pi}{2}}\,(Cl_{im}n_0)^{1/4}\left(1 - \frac{p_{min}}{p_{max}}\right)\right]^{-1}; \qquad (12.38)$$

and, consequently, in this case the optimal thickness distribution can be written as

$$h = \max\left\{h_0, \frac{aq_0}{2}\left(1 + \frac{b}{r}\right)\sqrt{\frac{\pi}{2}}\,(Cl_{im}n_0)^{\frac{1}{4}}\left(1 - \frac{p_{min}}{p_{max}}\right)\right\}.$$

12.3 Shape Optimization of Axisymmetric Shells

Consider shells with constant thickness h when $m = 4$ and

$$q_n = q_\varphi = 0, \quad 0 < x < L,$$

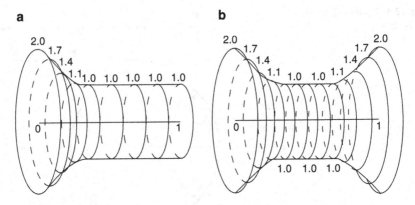

Fig. 12.5 Computational results: (a) $r(0) = 2$, $r(1) = 1$; (b) $r(0) = 2$, $r(1) = 2$

and external loads are applied to the free boundaries of the shell at $x = 0$ and $x = L$ with resultant force R (see Fig. 12.5). In this case

$$\sigma_\varphi = \frac{R}{2\pi rh}\chi(r), \quad \chi(r) = \sqrt{1 + \left(\frac{dr}{dx}\right)^2},$$

$$\sigma_\varphi = \frac{R}{2\pi h\chi(r)}\frac{d^2r}{dx^2} \tag{12.39}$$

and the optimization problem is formulated as

$$J = \int_0^L r\chi(r)dx \to \min \tag{12.40}$$

with constraints

$$\psi_1 = \frac{\beta}{r}\chi(r) - 1 \le 0, \tag{12.41}$$

$$\psi_2 = \frac{\beta}{\chi(r)}\left(\frac{d^2r}{dx^2}\right) - 1 \le 0, \tag{12.42}$$

$$\psi_3 = 1 - \frac{r}{r_g} \le 0, \tag{12.43}$$

where $R = (R)_{p=1}$ and the problem parameter β is defined as

$$\beta = \frac{R}{2\pi h\sigma_0}. \tag{12.44}$$

In what follows, we consider particular shape optimization problems of shell design when the thickness h is constant. Before starting with examples we make one note. For an unclosed shell supported usually by some rings or other arrangements against circumferential extension, some bending will occur near the support. But the edge effect is very localized and the edge zone with meaningful bending moments is relatively small. At a certain distance from the boundary we can use the membrane theory with satisfactory accuracy (see [TW59]). Taking this into account, we will construct the optimal design for such cases up to the small zone near the boundary.

To find the solution of the problem (12.40)–(12.44), we will use the method of penalty functions in combination with a genetic algorithm. To apply the penalty function method, we introduce the functions Ψ_i, $i = 1, 2, 3$, and penalty functionals J_i, $i = 1, 2, 3$,

$$\Psi_i = \begin{cases} \psi_i \text{ if } \psi_i > 0, \\ 0 \text{ if } \psi_i \leq 0, \end{cases} \tag{12.45}$$

$$J_i = \int_0^1 \Psi_i \, dx \tag{12.46}$$

and construct an augmented functional

$$J^a = J + \sum_{i=1}^3 \mu_i J_i. \tag{12.47}$$

Here $\mu_i \geq 0$ are the arbitrary positive parameters of the method.

For the minimization of the functional (12.47), which was built in the framework of the guaranteed approach, we apply the numerical optimization method, namely the genetic algorithm (GA) (see [Hol75, Gol89, HM03, BS93, MMM99]). Note also that another evolutionary algorithm can be applied here, see, for example, [HGK90, Ing93]. In the case under consideration this method (GA) appears to be effective in finding the global optimum. We divide the dimensionless length $L = 1$ of the shell into several segments of equal length, which define dimensionless x-coordinate of control points. There are k control points

$$(x_0, r(x_0)), (x_1, r(x_1)), \dots, (x_{n-1}, r(x_{n-1})),$$

$n - 2$ of them are free, and the first and the last points are fixed.

To find the global optimum, we shall consider a sequence of generations of population $S(t)$, where $t = 1, 2, \dots$ is the generation counter. This population $S(t)$ consists of M individuals P^j $(j = 0, \dots, M - 1)$

$$S(t) = \left(P^0(t), P^1(t), \dots, P^{M-1}(t) \right) \tag{12.48}$$

The number M of individuals in the population is assumed to be constant for all generations. Each individual $P^j(t)$ is the possible solution of the optimization problem and is described as

$$P^j(t) = \left((r(x_0))_t^j, (r(x_1))_t^j, \ldots, (r(x_{k-1}))_t^j \right). \qquad (12.49)$$

We suppose that nodes $(r(x_i))_t^j$ satisfy given constraints

$$r_g \leq (r(x_i))_t^j \leq r_{max},$$
$$i = 0, 1, \ldots, k-1, \quad j = 0, 1, \ldots, M-1, \qquad (12.50)$$
$$r_{max} = \max(r(0), r(1)).$$

Each node $(r(x_i))_t^j$ is represented by the string using the alphabet $A = (0, 1, 2, 3, 4, 5, 6, 7, 8, 9)$, that is

$$(r(x_i))_t^j \rightarrow \left(a_{i0}^j(t), a_{i1}^j(t), \ldots, a_{im}^j(t) \right). \qquad (12.51)$$

Here $a_{is}^j \in A$. For the nodes $(r(x_i))_t^j$ we have

$$(r(x_i))_t^j = r_g + \left(r_{max} - r_g \right) \frac{\sum_{s=0}^{m} 10^s * a_{is}^j(t)}{10^m - 1}, \qquad (12.52)$$

where the parameter $m > 0$ is taken to control the accuracy of computations.

The optimization problem is now expressed as

$$J^a(P^j(t)) \rightarrow \min. \qquad (12.53)$$

An important concept of the present algorithm is the concept of fixation. This means that some (i) randomly selected nodes of individual are fixed during several (F_{st}) generations, while the remaining ($k - i$) nodes vary. All individuals $P^j(0)$ in the initial population $S(0)$ are randomly generated. Initializations are repeated periodically for each $F_{st}(k - 1)$ generation.

The other important "genetic" features of the numerical method are the concepts of mutation, crossover and elitism. Mutation of the individual is realized by the choice of k stochastic numbers $j_0, j_1, \ldots, j_{k-1}$ uniformly distributed from 0 to m. Mutation operation changes each of the values $a_{0j_0}^j, a_{1j_1}^j, \ldots, a_{k-1j_{k-1}}^j$ by the new stochastic value belonging to the alphabet A. Thus, all nodes of the individual are mutated simultaneously, "The roulette wheel selection" is used for the choice of parent individuals for crossover. The crossover of two strings (12.51) is realized by the choice of k stochastic numbers $j_0, j_1, \ldots, j_{k-1}$ that possess a uniform probability distribution function. In two-point crossover, two crossing sites are selected as random (with the help of k stochastic numbers already described) and parent individuals exchange the segment that lies between two crossing sites.

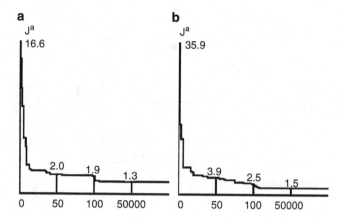

Fig. 12.6 The functional J^a dependence on number of generation (iteration)

The elitism selection method ensures that the best individual of the population must be copied from the current population $S(t)$ to the next population $S(t + 1)$. Elitism guarantees no increasing of the functional value.

Computations have been performed for the following values of parameters: $L = 1$, $r_g = 0.1$, $r(0) = 2.0$, 1.5, 1.0, 0.5, $r(1) = r_g$, $r(0)/2$, $r(0)$ ($r(0)$ and $r(1)$ are the left and right radius of the shell). Parameters of GA were taken as the number of individuals for population $M = 10$, the number of individuals nodes $k = 13$, and the fixation step $F_{st} = 20$, $m = 8$. Some computational results are presented in Fig. 12.5. The dependence of the functional J^a (12.47) on the number of generations (iterations) is shown in Fig. 12.6 for the cases (a and b), respectively in Fig. 12.5.

Considered in this section is the problem of optimal shell design based on fracture mechanics of brittle and quasi-brittle bodies, when the radius $r(x)$ is considered as a desired design variables and the thickness is supposed to be constant and given.

12.4 Simultaneous Optimization of the Meridian Shape and the Thickness Distribution of the Shell

In what follows, we assume that the external loads are applied to the free boundaries of the shell at $x = 0$ and $x = L$ (see Fig. 12.7).

These loads act parallel to the axis x in opposite directions, and their resultant forces are denoted by R and are supposed to be given. It is also assumed that the external forces are cyclic and vary quasi-statically within given limits, i.e.

$$R = pR^0, \quad 0 \le p_{min} \le p \le p_{max}, \tag{12.54}$$

where p is a loading parameter and p_{min}, p_{max} and R^0 are given values.

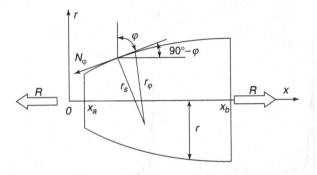

Fig. 12.7 Shell under loading

The optimization problem consists in finding the shape $r = r(x)$ and the thickness distribution $h = h(x)$ of the shell such that the volume of the shell material J (12.27) is minimized while satisfying the longevity constraint (12.10), the boundary conditions for the function $r(x)$

$$r(0) = r_1, \quad r(L) = r_2 \tag{12.55}$$

and the additional geometric isoperimetric constraint on the volume of the shell (12.28), where r_1, r_2, L and V_0 are given positive problem parameters.

12.4.1 Optimum Shells of Positive Gaussian Curvature

Now, we will suppose that the optimized shell is convex, i.e.

$$d^2r/dx^2 \leq 0 \tag{12.56}$$

(positive Gaussian curvature), and consider separately the cases $R^0 > 0$ (shell under tension) and $R^0 < 0$ (shell under compression). At first we assume that $R^0 > 0$. In this case, as is seen from the expressions

$$\sigma_\varphi = \frac{R^0}{2\pi r h} \chi(r),$$

$$\sigma_\theta = \frac{R^0}{2\pi h \chi(r)} \frac{d^2r}{dx^2}, \tag{12.57}$$

we have

$$\sigma_\varphi \geq 0, \quad \sigma_\theta \leq 0$$

and, consequently, the condition

$$\max\{\sigma_\varphi, \sigma_\theta\} \leq \sigma_0 \tag{12.58}$$

(see also (12.25), (12.26)) is written as

$$\frac{R^0}{2\pi r h}\chi(r) \leq \sigma_0. \tag{12.59}$$

It can be shown (see Section 12.4.3) that for an optimal solution of the problem (12.27), (12.28), (12.55), (12.58) equality is realized in (12.59), and, consequently, we obtain the following expression for the optimized thickness distribution:

$$h = \frac{R^0}{2\pi r \sigma_0}\chi(r). \tag{12.60}$$

Taking into account the expression (12.60), we construct the augmented Lagrange functional for the optimization problem. We have

$$J^a = J - \lambda V = \int_0^L \left[\frac{R^0}{\sigma_0}\chi^2(r) - \lambda \pi r^2\right]dx, \tag{12.61}$$

where λ is the Lagrange multiplier corresponding to the isoperimetric condition (12.28). The necessary extremum condition (Euler equation) for the functional (12.61) can be written as

$$\frac{d^2 r}{dx^2} + \frac{\lambda \pi \sigma_0}{R_0} = 0, \quad 0 \leq x \leq L. \tag{12.62}$$

The positiveness of the Lagrange multiplier

$$\lambda = -\frac{R^0}{\pi r \sigma_0}\frac{d^2 r}{dx^2} \geq 0$$

gives us the possibility to represent the general solution of the boundary-value problem (12.55), (12.62) in the following form:

$$r(x) = A \sin \nu x + B \cos \nu x, \quad \nu = \sqrt{\tilde{\lambda}}, \quad \tilde{\lambda} = \lambda \frac{\pi \sigma_0}{R^0}. \tag{12.63}$$

The constants A and B are found with the help of the boundary conditions (12.55). For simplicity, we present here the corresponding expressions for the symmetrical case ($r_1 = r_2$)

$$A = r_1 tg\left(\frac{\nu L}{2}\right), \quad B = r_1. \tag{12.64}$$

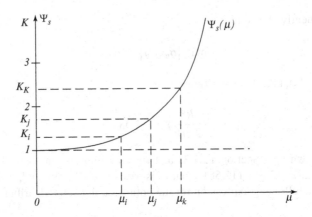

Fig. 12.8 $\Psi_s(\mu)$ function

Then we use (12.28), (12.63) and (12.64) to derive the algebraic equation

$$\kappa \equiv \frac{V_0}{\pi r_1^2 L} = \Psi_s(\mu), \quad \mu = \nu L,$$

$$\Psi_s(\mu) = \frac{1}{2}\left(tg^2\frac{\mu}{2} - 1\right)\left(1 - \frac{\sin\mu}{\mu}\right) - \frac{1}{\mu}(\cos\mu - 1)\,tg\frac{\mu}{2} + 1 \qquad (12.65)$$

for finding the parameter μ, which determines the values of λ and ν. The function Ψ_s is shown in Fig. 12.8.

The corresponding optimal shape of the shell $r = r(x)$ and its thickness distribution are expressed in the following form:

$$r(x) = r_1\left\{\cos\nu x + \frac{1 - \cos\nu L}{\sin\nu L}\sin\nu x\right\}, \qquad (12.66)$$

$$h = h(x) = \frac{R^0}{2\pi\sigma_0}T(x)$$

$$T(x) \equiv \frac{\left(1 + \tilde{\lambda}\left(A^2\cos^2\nu x + B^2\sin^2\nu x - AB\sin^2(2\nu x)\right)\right)^{1/2}}{A\sin\nu x + B\cos\nu x} \qquad (12.67)$$

and shown in Fig. 12.9 by the curves $r(x)$ and $h(x)$.

For the presented curves, the corresponding problem parameters have been taken as $L = 500$ cm, $r_1 = r_2 = 50$ cm, $R_0 = 10^6$ kN, $\sigma_0 = 10^4$ N/cm^2 and $V_0 = 7.715 \times 10^7$ cm^3.

Now consider the second case, when $R^0 < 0$ (shell under compression). In this case, as is seen from the formulas (12.57), we have

$$\sigma_\varphi \leq 0, \quad \sigma_\theta \geq 0$$

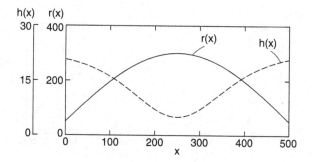

Fig. 12.9 Distribution of radius $r(x)$ and thickness $h(x)$ in the case $r_1 = r_2$ and $V_0 > \pi r_1 L$ (symmetrical case)

and, consequently, the constraint (12.58) takes the following form:

$$\frac{R^0}{2\pi h \chi(r)} \frac{d^2 r}{dx^2} \le \sigma_0. \qquad (12.68)$$

As in the previous case $(R^0 > 0)$, it can be shown that the equality is realized in (12.68) for the optimal design $(r(x), h(x))$. Thus, we have

$$h = \frac{R^0}{2\pi \sigma_0 \chi(r)} \frac{d^2 r}{dx^2}. \qquad (12.69)$$

Using Eqs. (12.27), (12.68) and (12.69), we derive the following expression for the augmented Lagrange functional:

$$J^a = J - \lambda V = \int\limits_0^L \left[\frac{R^0}{\sigma_0} r \frac{d^2 r}{dx^2} - \lambda \pi r^2 \right] dx. \qquad (12.70)$$

The necessary optimality condition is written as

$$\frac{d^2 r}{dx^2} + \widehat{\lambda} r = 0, \quad \widehat{\lambda} = -\lambda \frac{\pi \sigma_0}{R^0} \qquad (12.71)$$

for the considered case when

$$R^0 < 0, \quad d^2 r / dx^2 \le 0.$$

The Lagrange multiplier λ and the introduced parameter $\widehat{\lambda}$ will be positive

$$\lambda = \frac{R^0}{\pi r \sigma_0} \frac{d^2 r}{dx^2} \ge 0, \quad \widehat{\lambda} \ge 0.$$

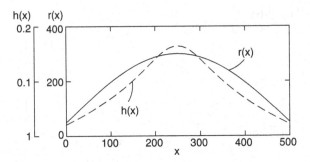

Fig. 12.10 Distribution of radius $r(x)$ and thickness $h(x)$ in the case $r_1 = r_2$ and $V_0 > \pi r_1 L$ (symmetrical case)

To find the optimal shape of the meridian $r = r(x)$, we use the boundary-value problem (12.55), (12.71), which coincides with the boundary-value problem (12.55), (12.62) considered in the previous case, $R^0 \geq 0$. Consequently, the shape of the meridian $r = r(x)$ will be given by expression (12.63), when the unknown constants A, B and $\tilde{\lambda}$ are determined with the help of formulas (12.64) and (12.65), corresponding to the symmetrical case $r_1 = r_2$. Thus, the optimal shape of the meridian $r = r(x)$ obtained for the case of the shell under compression is the same as in the previous case of the shell under tension $(R^0 \geq 0)$. Despite the coincidence of the optimal meridian shapes for tensioned and compressed shells, the corresponding thickness distributions differ from one another. For the shell under compression, we will have the following expression for the thickness distribution:

$$ h = -\frac{R^0}{2\pi\sigma_0}\frac{\widehat{\lambda}}{T(x)}, \tag{12.72} $$

where the function $T(x)$ is defined by the formula (12.67).

It is seen in (12.72) that the thickness is distributed along the axis x in an opposite manner as compared with the previous case of a tensioned shell. The distribution of the design variables is shown in Fig. 12.10 by the curves $r(x)$ and $h(x)$ and corresponds to the following value of the parameters: $L = 500\,\text{cm}$ $r_1 = r_2 = 50\,\text{cm}$, $R_0 = -10^6\,\text{kN}$, $\sigma_0 = 10^4\text{N/cm}$, $V_0 = 7.715 \times 10^7\text{cm}^3$.

12.4.2 Optimum Shells of Negative Gaussian Curvature

Consider the case when the optimized shell is concave, i.e. Gaussian curvature is negative

$$ d^2r/dx^2 \geq 0 $$

and suppose that the shell is under tension, i.e. $R^0 > 0$. In this case, as is seen from the expressions (12.57), we have

$$\sigma_\varphi \geq 0, \quad \sigma_\theta \geq 0$$

and, consequently, the condition (12.58) can be written as

$$\sigma_\theta \leq \sigma_\varphi \leq \sigma_0$$

or as

$$\sigma_\varphi \leq \sigma_\theta \leq \sigma_0.$$

Taking into account that for optimal solution of the problem (12.27), (12.28), (12.55) and (12.58), the equality is realized (12.58) (see Section 12.4.3); we can rewrite the equality as

$$\sigma_\varphi = \sigma_0, \quad \sigma_\theta \leq \sigma_\varphi, \tag{12.73}$$

$$\sigma_\theta = \sigma_0, \quad \sigma_\varphi \leq \sigma_\theta. \tag{12.74}$$

Consider at first the case (12.73); we can use the expression (12.60) for the thickness distribution, and the expression (12.61) for the augmented Lagrange functional. The necessary extremum condition can be written as

$$\frac{d^2 r}{dx^2} - \gamma_1 r = 0, \quad 0 \leq x \leq L,$$

$$\gamma_1 = -\frac{\lambda \pi \sigma_0}{R^0} \geq 0. \tag{12.75}$$

Taking into account the positiveness of the parameter γ_1, we find the shape of the shell, which satisfies (12.75) and the boundary conditions (12.55)

$$r(x) = A_1 e^{sx} + A_2 e^{-sx}, \quad s = \sqrt{\gamma_1},$$

$$A_1 = \frac{r_2 - r_1 e^{-sL}}{e^{sL} - e^{-sL}}, \quad A_2 = r_2 - A_1. \tag{12.76}$$

The corresponding optimal thickness distribution is given by the following expression:

$$h(x) = \frac{R^0 \Phi_1(x)}{2\pi\sigma_0}, \quad 0 \leq x \leq L, \tag{12.77}$$

$$\Phi_1 = \frac{\sqrt{1 + s^2 \left(A_1^2 e^{2sx} + A_2^2 e^{-2sx} - 2A_1 A_2\right)}}{A_1 e^{sx} + A_2 e^{-sx}}, \tag{12.78}$$

where $s = \sqrt{\gamma_1}$. The value of the parameter γ_1 can be found with the help of the isoperimetric condition (12.28). The solution is realized if

$$\sigma_\theta \leq \sigma_\varphi$$

with the help of the expression (12.57) for σ_θ, σ_φ and (12.45), the last inequality is transformed to

$$r\frac{d^2r}{dx^2} \leq 1 + \left(\frac{dr}{dx}\right)^2. \tag{12.79}$$

Using the inequality (12.79) and the representation (12.76), we find the following condition for the existence of the considered type of solution:

$$\gamma_1 = \frac{1}{4A_1A_2}. \tag{12.80}$$

Consider now another case when relations (12.44) are realized. Using the expressions (12.60), (12.61) for the thickness distribution and for augmented Lagrange functional, we derive the Euler equation

$$\frac{d^2r}{dx^2} - \gamma_2 r = 0, \quad 0 \leq x \leq L,$$

$$\gamma_2 = \frac{\lambda\pi\sigma_0}{R^0} \geq 0 \tag{12.81}$$

and find the optimal shape which satisfies (12.81) and the boundary conditions (12.55). Taking into account the positiveness of the parameter γ_2, we construct the solution in the form (12.76) with $s = \sqrt{\gamma_2}$. The corresponding optimal thickness distribution can be represented in the following form:

$$h(x) = \frac{R^0\Phi_2(x)}{2\pi\sigma_0}, \tag{12.82}$$

$$\Phi_2(x) = \frac{s^2}{\Phi_1(x)}, \quad s = \sqrt{\gamma_2}. \tag{12.83}$$

The value of γ_2 is found with the help of the isoperimetric condition (12.28). The considered type of solution exists if $\sigma_\varphi \leq \sigma_\theta$. This inequality is transformed to the following form

$$r\frac{d^2r}{dx^2} \geq 1 + \left(\frac{dr}{dx}\right)^2. \tag{12.84}$$

If we substitute the obtained solution for $r = r(x)$ into relation (12.81), we will have the condition

$$\gamma_2 \geq \frac{1}{4A_1A_2}$$

of existence of the investigated solution. Here A_1 and A_2 are considered as functions of the parameter γ_2.

12.4.3 Some Properties of Optimal Solution

To find the optimal solution of the problem (12.27 (12.10), (12.28), (12.55)), we used the relation (12.60) which is a consequence of equality in (12.59). We shall now prove that for the optimal solution, only this case must be realized; that is, there is no part of the optimal shell where the rigorous *inequality* is satisfied in (12.59). Let us assume the contrary; that is, we assume that at some segment $[x_1, x_2]$ $(0 \leq x_1 \leq x_2 \leq L)$, the optimal solution $(r^*(x), h^*(x))$ satisfies the rigorous inequality in (12.59) or

$$h^* > \frac{R_0}{2\pi r^* \sigma_0} \chi(r^*).$$

For the other intervals considered $(0 \leq x < x_1$ and $x_2 < x \leq L)$, the optimal solution $(r^*(x), h^*(x))$ is supposed to satisfy relation (12.60) (rigorous inequality is realized in (12.59)). Then, we can construct a new admissible design $\left(\widehat{r}(x), \widehat{h}(x)\right)$ in the following form:

$$\widehat{r} = r^*, \quad \widehat{h} = h^* \quad \text{if } 0 \leq x \leq x_1 \text{ and } x_2 < x \leq L,$$
$$\widehat{r} = r^*, \quad \widehat{h} = \frac{R^0}{2\pi r^* \sigma_0} \chi(r^*) < h^*, \quad \text{if } x_1 \leq x \leq x_2, \qquad (12.85)$$

which satisfies the strength constraint (12.59) and the isoperimetric condition (12.28), i.e. $\left(\widehat{r}, \widehat{h}\right)$ is an admissible design. Note that for the constructed admissible design $\left(\widehat{r}, \widehat{h}\right)$, the rigorous equality is realized in (12.59) for the total interval $0 \leq x \leq L$. Thus, the constructed admissible solution is the equal-strength design, and for this solution, we will have

$$J(\widehat{r}, \widehat{h}) = 2\pi \int_0^{x_1} r^* h^* \chi(r^*) dx + 2\pi \int_{x_1}^{x_2} r^* \widehat{h} \chi(r^*) dx +$$

$$+2\pi \int_{x_2}^{L} r^* h^* \chi(r^*) dx < 2\pi \int_0^{L} r^* h^* \chi(r^*) dx = J(r^*, h^*).$$

The contradiction

$$J(r^*, h^*) > J(\widehat{r}, \widehat{h})$$

proves the assumption that for optimal design, the rigorous equality in (12.59) must be realized for the total interval $[0, L]$. In analogous manner, it can be shown that the rigorous equality is realized in (12.68) for the optimal design $(r(x), h(x))$, $0 \leq x \leq L$, in the case $R^0 < 0$.

Chapter 13
Uncertainties in Material Characteristics

13.1 Discrete Sets of Materials with Uncertain Properties

13.1.1 Basic Representations

This section deals with problems of optimal design of structures from various materials. The number of materials is supposed to be finite and consequently the admissible design set consists of separate discrete values. Suppose that material i $(i = 1, 2, \ldots, r)$ is characterized by the following property vector (see Fig. 13.1):

$$\xi_i = \{\xi_i^1, \xi_i^2, \ldots, \xi_i^m\}, \quad i = 1, 2, \ldots, r, \tag{13.1}$$

where r is the number of given materials (steel, titanium, ...) and m is the number of material properties essential for the problem (material density, Young's modulus, ...).

Taking into account that each of the given materials can be enumerated with one parameter, we apply natural parameterization using the scalar variable t that takes given values $t_1, t_2, \ldots, t_r$, i.e.

$$\xi(t_i) = \xi_i, \quad t \in \{t_1, t_2, \ldots, t_r\}. \tag{13.2}$$

For example, if the materials are characterized by Young's modulus E, linear expansion coefficient α and material density ρ then

$$E(t_i) = E_i, \quad \alpha(t_i) = \alpha_i, \quad \rho(t_i) = \rho_i.$$

We assume that there is incomplete information concerning the material properties, and it is possible to represent all material characteristics in the form

$$\xi_i^j = \widehat{\xi}_i^j + \eta_i^j \tag{13.3}$$

$$-\Delta_i^j \le \eta_i^j \le \Delta_i^j \tag{13.4}$$

N.V. Banichuk and P.J. Neittaanmäki, *Structural Optimization with Uncertainties*, Solid Mechanics and Its Applications 162, DOI 10.1007/978-90-481-2518-0_1,
© Springer Science+Business Media B.V. 2010

Fig. 13.1 Considered
materials and their properties

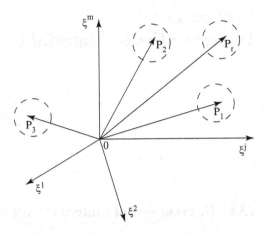

where $\widehat{\xi}_i^j$ ($i = 1, 2, \ldots, r$, $j = 1, 2, \ldots, m$) are the given characteristics, Δ_i^j are given positive constants, and η_i^j are unknown variables in (13.3), (13.4). For material i there is a set Λ_{η_i} of admissible vectors $\eta_i = \{\eta_i^1, \eta_i^2, \ldots, \eta_i^m\}$, i.e.

$$\eta_i \in \Lambda_{\eta_i}, \ i = 1, 2, \ldots, r. \tag{13.5}$$

In what follows, we consider a one-dimensional problem of optimal design of a rod from a discrete set of materials, taking into account the incomplete information concerning the material properties. Let the elastic rod lie along the x-axis ($0 \le x \le L$) and be fixed at the ends $x = 0$ and $x = L$. The ends are fixed:

$$u(0) = u(L) = 0, \tag{13.6}$$

where $u(x)$ is the displacement distribution along the x-axis. The rod consists of a discrete set of materials, distributed along the x-axis and characterized by the set of parameters

$$\{E_i, \alpha_i, \rho_i\}, \ i = 1, 2, \ldots, r,$$

where E_i is Young's modulus, α_i is the linear expansion coefficient, ρ_i is the material density, i is the material index, and r is the number of materials. Distribution of the parameters $(E(x), \alpha(x), \rho(x))$ along the rod are given by piece-wise constant functions defined on the segment $0 \le x \le l$. For each point $x \in [0, L]$ these functions take some values from the given finite set, i.e.

$$E(x) \in \{E_i\}, \quad \alpha(x) \in \{\alpha_i\}, \quad \rho(x) \in \{\rho_i\},$$

where $i = 1, 2, \ldots, r$. In what follows, we apply parameterization (Fig. 13.2) of the essential parameters using the piece-wise constant function $t = t(x)$ ($x \in [0, L]$) taking the value $t = t_i = i$ from the given set, i.e. $t \in \{t_i = i\}$ and such that the following equalities are satisfied:

Fig. 13.2 Piece-wise
constant distribution of
material properties along the
structure

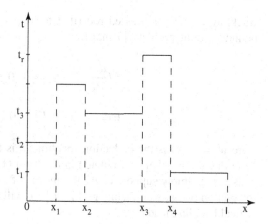

$$E(t(x))_{t=t_i=i} = E_i, \ \alpha(t(x))_{t=t_i=i} = \alpha_i, \ \rho(t(x))_{t=t_i=i} = \rho_i. \quad (13.7)$$

Suppose that the rod is in an undeformed (natural) state for the initial temperature T. Heating of the rod is performed up to the temperature $T + \theta\varphi(x)$. The total strain ε is a sum of elastic ε_e and thermal ε_t strains, i.e. $\varepsilon = \varepsilon_e + \varepsilon_t$, where $\varepsilon_t = L\Delta T = L\theta\varphi(x)$. The generalized Hooke's law gives the following relation between stress σ and strain ε [Lov44, Tim87, Now70]

$$\sigma = E(\varepsilon - \alpha\theta\varphi(x)). \quad (13.8)$$

Taking into account that

$$\sigma = -\frac{P}{S}, \quad \varepsilon = \frac{du}{dx},$$

we will have

$$u(L) - u(0) = -P \int_0^L \frac{dx}{SE} + \theta \int_0^L \varphi\alpha \, dx, \quad (13.9)$$

where S is the cross-section area of the rod and $P > 0$ is the compressive force (clamping reaction), acting on the inhomogeneous elastic rod.

As a result of the considered clamping conditions (13.6) and the relation (13.9) we will have the following value for compressed force:

$$P = \theta\frac{J_1}{J_2}, \quad (13.10)$$

$$J_1 = \int_0^L \varphi \, \alpha \, dx, \quad J_2 = \int_0^L \frac{dx}{SE}. \quad (13.11)$$

Buckling of the compressed rod (if it takes place) is described by the following boundary value problem [Tim56]:

$$EI\frac{d^2w}{dx^2} + Pw = 0, \ 0 \le x \le L \tag{13.12}$$

$$w(0) = 0, \ w(L) = 0 \tag{13.13}$$

Here $w = w(x)$ is the deflection function, I is the moment of inertia of the rod. Note that the problem, formulated in the form (13.12), (13.13), corresponds to the ends being simply supported. By the means of multiplication of (13.12) by w and integration by parts on the interval $[0, L]$ we will have the following expression for critical buckling load

$$P = \frac{J_3}{J_4}, \tag{13.14}$$

$$J_3 = \int_0^L \left(\frac{dw}{dx}\right)^2 dx, \ J_4 = \int_0^L \frac{w^2}{EI}dx. \tag{13.15}$$

Using Eqs. (13.10), (13.11), (13.14), (13.15) for buckling load and reaction force arising from the heating process, we obtain the critical heating temperature

$$\theta = P\frac{J_2}{J_1} = \frac{J_2 J_3}{J_1 J_4}. \tag{13.16}$$

Note that to find the buckling load P and the corresponding buckling mode $w(x)$ we can use the well-known variational principle [Was82, CB73, Kom88a, Kom88b]

$$P = \min_w \frac{J_3}{J_4}. \tag{13.17}$$

The minimum in (13.17) is found on the set of admissible functions satisfying boundary conditions (13.13).

Suppose that the temperature increment is a given positive function and the rod has cylindrical shape, i.e.

$$\Delta T = \theta\varphi(x) > 0, \ S(x) = S_0 > 0, \tag{13.18}$$

where S_0 is a given value. Suppose also that the densities of materials are known, but the Young's modulus E and the linear expansion coefficient are given with incomplete information, i.e.

$$E = \widehat{E} + \eta^E, \ \alpha = \widehat{\alpha} + \eta^\alpha, \tag{13.19}$$

$$\eta = \left\{ \eta^E, \eta^\alpha \right\}, \tag{13.20}$$

$$\Lambda_\eta = \left\{ -\Delta^E \le \eta^E \le \Delta^E, -\Delta^\alpha \le \eta^\alpha \le \Delta^\alpha \right\}, \tag{13.21}$$

where Δ^E and Δ^α are given problem parameters.

13.1.2 Minimization of Reaction Force

Using a worst case scenario, consider minimizing the reaction force P under mass constraint

$$P_{\min} = \theta \min_t \max_{\eta \in \Lambda_\eta} \frac{J_1(t, \eta^\alpha)}{J_2(t, \eta^E)}, \tag{13.22}$$

$$J_1(t, \eta^\alpha) = \int_0^L \left[\widehat{\alpha}\left(t(x)\right) + \eta^\alpha(x) \right] \varphi(x) dx, \tag{13.23}$$

$$J_2(t, \eta^E) = \frac{1}{S_0} \int_0^L \frac{dx}{\widehat{E}(t(x)) + \eta^E(x)}, \tag{13.24}$$

$$M(t) = S_0 \int_0^L \rho(t(x)) dx \le M_0, \tag{13.25}$$

$$t(x) = \{t_i | x_{k-1} < x < x_k, \quad t = 1, 2, \ldots, n, \quad x_0 = 0, \quad x_n = L\}. \tag{13.26}$$

Here $M_0 > 0$ is a given constant. Thus, we formulate the problem of optimization of the quality functional P (13.22)–(13.24) with the mass constraint (13.25) using the piece-wise constant function $t(x)$ (13.26) as a design variable, characterizing the structure of the inhomogeneous rod (number of pieces, sizes of pieces, material properties). Values of t_i belong to the given discrete set $t_i \in \{1, 2, \ldots, r\}$. If the mass constraint (13.25) is not taken into account or is not active and the temperature distribution is homogeneous, i.e. $\varphi(x) = \varphi_0 > 0$ (φ_0 is a given constant), then for arbitrary piece-wise constant distribution of material properties the optimized functional can be estimated as

$$P = \theta \frac{J_1}{J_2} = \theta \varphi_0 S_0 \left\{ \sum_{k=1}^n \left(\widehat{\alpha}_k + \eta_k^\alpha \right) (\Delta l)_k \right\} \left\{ \sum_{k=1}^n \frac{(\Delta l)_k}{\left(\widehat{E}_k + \Delta_k^E \right)} \right\}^{-1}$$

$$\le \theta \varphi_0 S_0 \left\{ \sum_{k=1}^n \left(\widehat{\alpha}_k + \Delta_k^\alpha \right) (\Delta l)_k \right\} \left\{ \sum_{k=1}^n \frac{(\Delta l)_k}{\left(\widehat{E}_k + \Delta_k^E \right)} \right\}^{-1}. \tag{13.27}$$

and, consequently, the internal maximum in (13.22) with respect to $\eta \in \Lambda_\eta$ is achieved when

$$\eta_k^\alpha = \Delta_k^\alpha, \ \eta_k^E = \Delta_k^E, \ k = 1, 2, \ldots, n. \tag{13.28}$$

Thus the optimal design problem with incomplete information is transformed to a conventional problem of reaction force minimization

$$P_{min} = \theta \varphi_0 S_0 \min_t \frac{\int_0^L [\widehat{a}(t(x)) + \Delta^\alpha] \, dx}{\int_0^L \frac{dx}{[\widehat{E}(t(x)) + \Delta^E]}} \tag{13.29}$$

under constraints (13.25), (13.26).

Note that when the mass constraint (13.25) is not taken into account or is not active the optimal solution of the problem (13.25), (13.26), (13.29) is realized for a homogeneous rod composed of material with minimal value β_1 of the parameter

$$\beta = \left(\widehat{E} + \Delta^E\right)\left(\widehat{a} + \Delta^\alpha\right).$$

We have

$$\beta_1 = \left\{\left(\widehat{E} + \Delta^E\right)\left(\widehat{a} + \Delta^\alpha\right)\right\} = \min_{1 \le i \le r} \left\{\left(\widehat{E_i} + \Delta_i^E\right)\left(\widehat{a}_i + \Delta_i^\alpha\right)\right\}. \tag{13.30}$$

To prove this, consider an arbitrary inhomogeneous rod consisting of n homogeneous parts of the length $(\Delta l)_k$ $(k = 1, 2, \ldots, n)$ with

$$\beta_1 = \left(\widehat{E}_1 + \Delta_1^E\right)\left(\widehat{a}_1 + \Delta_1^\alpha\right) < \beta_2 = \left(\widehat{E}_2 + \Delta_2^E\right)\left(\widehat{a}_2 + \Delta_2^\alpha\right) <$$
$$< \ldots < \beta_r = \left(\widehat{E}_r + \Delta_r^E\right)\left(\widehat{a}_r + \Delta_r^\alpha\right) \tag{13.31}$$

and perform the following estimations:

$$P = \theta \frac{J_1}{J_2} = \theta \varphi_0 S_0 \left\{\sum_{k=1}^n \left(\widehat{a}_k + \Delta_k^\alpha\right)(\Delta l)_k\right\} \left\{\sum_{k=1}^n \frac{(\Delta l)_k}{\left(\widehat{E}_k + \Delta_k^E\right)}\right\}^{-1} =$$
$$= \theta \varphi_0 S_0 \left\{\sum_{k=1}^n \frac{\beta_k}{\left(\widehat{E}_k + \Delta_k^E\right)}(\Delta l)_k\right\} \left\{\sum_{k=1}^n \frac{(\Delta l)_k}{\left(\widehat{E}_k + \Delta_k^E\right)}\right\}^{-1} \ge \beta_1 \theta \varphi_0 S_0 = P_1. \tag{13.32}$$

Thus we obtain the required inequality.

To solve the optimization problem (13.22)–(13.26) in general case, i.e. taking into account the isoperimetric inequality (13.25) and considering various temperature distributions $\theta\varphi(x)$, let us apply the approach (method of penalty functions), based on minimization of the augmented functional J^a defined by the formulae

$$J^a = P + \lambda\,(M - M_0) = \theta\frac{J_1}{J_2} + \lambda\,(M - M_0), \tag{13.33}$$

$$\lambda = \begin{cases} 0, & \text{if } M - M_0 \le 0, \\ \lambda_0 > 0, & \text{if } M - M_0 > 0, \end{cases} \tag{13.34}$$

where λ_0 is a positive penalty multiplier and the values η_k^α and η_k^E in the expression for functional P are taken as in (13.28).

Solution of the problem of the functional J^a (13.33), (13.34) minimization for various values of the problem parameter M_0 and material characteristics presented in Table 13.1 is performed with the help of a genetic algorithm [Gol89, Hol75, HGK90, HG92, HJK+00, HN96]. It is supposed that $L = 1$ and the interval $[0, 1]$ of the variable x is divided by the points x_i, $i = 1, 2, \ldots, n$, into $n - 1$ subintervals of equal length Δx. For each subinterval the values

$$\alpha = \widehat{\alpha} + \Delta^\alpha,\ E = \widehat{E} + \Delta^E,\ \rho$$

can take constant values corresponding to the chosen materials.

The index of the material can takes the values from 1 to 6. Populations under consideration consist of N individuals, represented admissible piece-wise homogeneous rods. The number N is supposed to be even and is kept constant in the population renewal process. Each j-individual of the population is described by the set of values $t(j, i)$, representing the design variable at a node. The "best" individual, i.e. the set $t(j, i)$ minimizing the augmented functional, is sought by using the genetic algorithm.

The first step of the algorithm consists in initialization of the population, that is assigning random values taken from $[1, \ldots, 6]$ to each element $t(j, i)$. For the created individuals ($j = 1, \ldots, N$) of the initial population we complete the augmented functional $J^a(j)$ and find the individual having the minimal value of the functional. Using the initial data and the next step of the algorithm, it is possible

Table 13.1 Properties of materials

N	Material	$(\widehat{E} + \Delta^E)$ $\times 10^{-5}$ MPa	$(\widehat{\alpha} + \Delta^\alpha)$ $\times 10^6$ (grad)$^{-1}$	$\rho g \times 10^4$ Nm^{-3}	$\mu \times 10$ MPa (grad)$^{-1}$
1	Molybdenum	3.3	5.6	10.2	18.48
2	Steel 1X17H2	2.0	10.3	7.75	20.6
3	Brass	1.1	19.0	8.6	20.9
4	Steel 30ХГСА	1.98	11.0	7.85	21.78
5	Steel 2X13	2.2	10.1	7.75	22.22
6	Steel 20X	2.07	11.3	7.74	23.391

to determine a new population consisting of N individuals, and so to successively minimize the functional J^a.

At the second step of the algorithm we select $N/2$ individual pairs, "parents", to obtain $N/2$ pairs of individuals, "children", that constitute new population. Selection of the first parent ("a") is performed by the following manner. Some natural number N^T is chosen and then N^T individuals are selected randomly. From this set of individuals we preserve and use only one individual having the minimal value of augmented functional J^a. Similarly we find the second parent ("b") and put together the first pair of individuals. All together we choose $N/2$ such pairs.

The third step of the algorithm consists in obtaining of two children from each pair of parents. For this purpose we take some constant value from the interval $[0, 1]$, that is called the crossover probability p_{co}. Then for each parent pair a random number p_r is selected from the interval $[0, 1]$, and a random natural number m from $[1, \ldots, n]$. If $p_r \leq p_{co}$ then the values of design variables of children at the nodes $i = 1, 2, \ldots, m$ are copied from their parents "a" and "b", but the meaning of these values at the nodes $i = m + 1, \ldots, n$ are obtained with the help of crossover. The latter means that for child "a" we copy the values in the corresponding nodes of the parent "b" and vice versa. Successive sorting of all parent pairs and performing of described operations lead to N individuals – children, that compose the new population.

The fourth step of the algorithm consists in mutation of the new population. This step is necessary to avoid staying at a local minimum of the functional. To realize the mutation procedure we take some small (~ 0.005) parameter p_m (probability of mutation). Then for all nodes of each individual of the population we generate a random number p_r from the interval $[0, 1]$. If $p_r \leq p_m$ then the value of design variable at this node is replaced by the arbitrary value, satisfying given constraints. The mutation procedure is not performed for the values $t(j, i)$ at the nodes with numbers $i = 1$ and $i = n$. For the new population, we compute the functionals $J^a(j)$ and select the best individual. Then we go to the second step of applied algorithm. Note that if the best child from the new population is worse then the best parent from the previous population then we replace it by this parent. This makes the process of finding of global minimum a monotonic one.

The optimal distribution of material $t(x)$ was determined with the help of this genetic algorithm. Here the parameters of computational process were taken as $n = 21, N = 10, N^T = 4, p_{CO} = 0.5, p_m = 0.05$. Calculations were completed after 75,000 generations. Characteristics of materials considered as admissible for optimal design of the nonhomogeneous rod are presented in Table 13.1. The results of numerical solution in the case of

$$\varphi(x) = 1, \ 0 \leq x \leq 1 \tag{13.35}$$

and for the following values of the problem parameter $W_0 = gM_0 = 7.75, 7.8, 8.0, 8.5, 9.0$ (g is acceleration of gravity) are presented in Table 13.2 and in Fig. 13.3. Variants (a)–(e) correspond to the values $W_0 = gM_0 = 12.0, 9.0, 8.5, 8.0, 7.8$.

Table 13.2 Determined optimal design

Considered cases	Constraint on weight $W_0 = M_0 g$	Weight of the rod $W = Mg$	Distribution of material along the rod (intervals)
1	7.75	7.75	2–20
2	7.8	7.75	2–20
3	8.0	7.995	1–2; 2–18
4	8.5	8.485	1–6; 2–14
5	9.0	8.975	1–10; 2–10

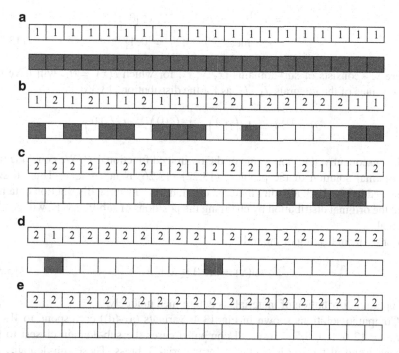

Fig. 13.3 Distribution of materials along the rod

Table 13.2 shows that the number of intervals occupied by molybdenum, the material having the minimum value of

$$\mu = \left(\widehat{E} + \Delta^E \right) \left(\widehat{\alpha} + \Delta^\alpha \right) = 18.48,$$

increases with increase of the problem parameter W_0. Figure 13.3 shows the computed distributions of materials. There are sub-domains occupied by molybdenum (dark grey color) and sub-domains occupied by steel 1X17H2 (white).

Note one important property of the optimal distribution of the materials for a uniform temperature distribution (13.35). Suppose that we have some admissible distribution $t = t^{(1)}(x)$ $(0 \leq x \leq l)$ of materials from (13.26)

$$t = t^{(1)}(x) = t_i, \quad x \in \Omega_i \subset \Omega,$$

$$l_i = \text{meas } \Omega_i, \quad \cup \Omega_i = [0, 1], \tag{13.36}$$

where Ω_i consists of sub-domains $[x_{k-1}, x_k]$ for which $t^{(1)} = t_i$. Then another admissible distribution $t = t^{(2)}(x)$ $(0 \leq x \leq 1)$ of the materials from (13.26)

$$t = t^{(2)}(x) = t_i, \ x \in \Omega'_i \subset [0, 1],$$

$$l_i = \text{meas } \Omega'_i, \ \cup \Omega'_i = [0, 1], \tag{13.37}$$

where Ω'_i consists of sub-domains $[x_{s-1}, x_s]$ for which $t(x) = t_i$, will have the same values of the integrals J_1, J_2, as for the distribution (13.36), i.e.

$$J_1\left(t^{(1)}\right) = J_1\left(t^{(2)}\right), \ J_2\left(t^{(1)}\right) = J_2\left(t^{(2)}\right)$$

and, consequently, the same value of the optimized functional P. We conclude that the optimal solution of the problem (13.22)–(13.26) is not unique. Thus, beside presented in Fig. 13.3, all other material distributions, material distribution obtained from the original distribution by changing the positions of sub-intervals, will be also optimal.

For uniform heating of the rod

$$\varphi(x) = x, \ 0 \leq x \leq 1, \tag{13.38}$$

the optimal distribution of materials is determined uniquely.

Computed solutions shown in Fig. 13.4 (variants (a)–(d)) correspond to $W_0 = gM_0 = 12.0, 9.0, 8.5, 8.0$. In all computed cases the sub-domain closest to the unheated end of the rod is occupied by material 3, brass. These sub-domains are shown by horizontal lines. Note that of the six materials, brass has the minimal elastic modulus.

Computations were also performed for the rod when the heating was given by

$$\varphi(x) = x(1 - x), \ 0 \leq x \leq 1 \tag{13.39}$$

and the results are shown in Fig. 13.5; variants (a)–(d) correspond to the following values of the parameter $W_0 = gM_0 = 12.0, 9.0, 8.5, 8.0$. The sub-domains occupied by brass in Fig. 13.5 are situated symmetrically with respect to the center of the rod.

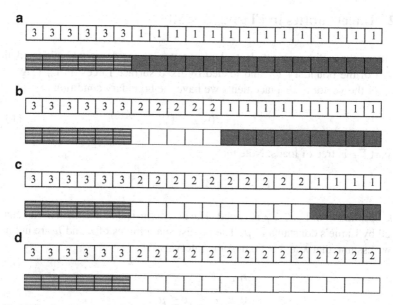

Fig. 13.4 Distribution of materials for $\varphi(x) = x, 0 \le x \le 1$

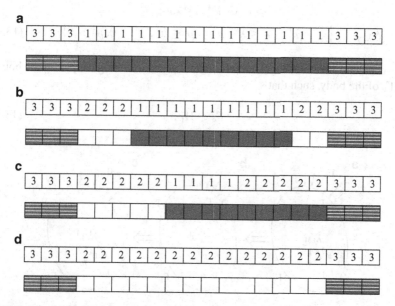

Fig. 13.5 Distribution of materials for $\varphi(x) = x(1 - x), 0 \le x \le 1$

13.2   Uncertainties in Elastic Moduli

Consider a three-dimensional elastic body occupying the domain Ω, clamped along the part of the boundary Γ_u, and loaded by fixed surface forces on Γ_σ (Fig. 13.6). Thus, for the vector of displacements we have the boundary condition

$$u = 0, \; x \in \Gamma_u. \tag{13.40}$$

The part Γ_f is free of loads. Note that

$$\Gamma_u + \Gamma_\sigma + \Gamma_f = \Gamma = \partial\Omega.$$

In the optimization, $\Gamma_v \subset \Gamma_f$ is varied. The body consists of elastic material characterized by Lamè's constants λ, μ. The precise magnitudes of λ and μ are unknown and we assume that λ and μ lie in some intervals

$$0 < \lambda^l \leq \lambda \leq \lambda^u,$$
$$0 < \mu^l \leq \mu \leq \mu^u,$$

where lower limits λ^l, λ^l and upper limits λ^u, μ^u are given. Thus

$$(\lambda, \mu) \in \Lambda_{\lambda,\mu},$$
$$\Lambda_{\lambda, \mu} = \left\{ \lambda, \mu | \lambda^l \leq \lambda \leq \lambda^u, \; \mu^l \leq \mu \leq \mu^u, \; \lambda^l > 0, \mu^l > 0 \right\}. \tag{13.41}$$

The shape optimization, considered below, consists of finding the part of the boundary Γ_v of the body, such that

$$\Gamma_v \subset H_\Gamma, \tag{13.42}$$

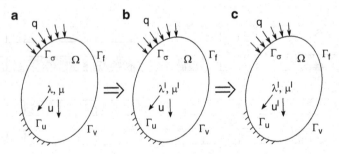

Fig. 13.6 (a) Equilibrium displacement field u for elastic moduli λ, μ; (b) Admissible displacement field u for elastic moduli λ^1, μ^1; (c) Equilibrium displacement field u^1 for elastic moduli λ^1, μ^1

where H_Γ is a given set of admissible surfaces and

$$J = \frac{1}{2} \langle q, u \rangle = \frac{1}{2} \int_{\Gamma_\sigma} q u d\Gamma_\sigma \rightarrow \min_{\Gamma_\nu \in H_\Gamma} \qquad (13.43)$$

under some geometric and strength constraints imposed on the displacements u and stresses fields. Taking into account (13.41) and a worst case scenario, we formulate the optimization problem with incomplete information as

$$J_* = \min_{\Gamma_\nu \subset H_\Gamma} \max_{(\lambda, \mu) \in \Lambda_{\lambda, \mu}} \frac{1}{2} \langle q, u \rangle. \qquad (13.44)$$

For arbitrary admissible Γ_ν satisfying (13.42), the maximum in (13.44) is realized when

$$\lambda = \lambda^l, \ \mu = \mu^l \qquad (13.45)$$

To prove this assertion, let us look at the total potential energy of the body at equilibrium:

$$W(u) = \int_\Omega \Pi d\Omega - \int_{\Gamma_\sigma} q u d\Gamma_\sigma. \qquad (13.46)$$

Here u denotes the displacement field corresponding to the equilibrium state, and Π is the potential energy density for the deformed elastic body:

$$\Pi = \frac{1}{2} \lambda (\varepsilon_{11} + \varepsilon_{22} + \varepsilon_{33})^2 + 2\mu (\varepsilon_{11}^2 + \varepsilon_{22}^2 + \varepsilon_{33}^2 + 2\varepsilon_{12}^2 + 2\varepsilon_{23}^2 + 2\varepsilon_{13}^2). \qquad (13.47)$$

Small elastic strains ε_{ij} are written as

$$\varepsilon_{ij} = \frac{1}{2} \left(\frac{\partial u_i}{\partial x_j} + \frac{\partial u_j}{\partial x_i} \right), \ i, j = 1, 2, 3. \qquad (13.48)$$

Let us compare an elastic body having some admissible λ and μ from (13.41) and an altered body with reduced magnitudes of the elastic constants (13.45) (in some subdomain $\Omega_l(\Omega)$ and with the same displacement field). Equations (13.47) and (13.48) show that

$$W(u) \geq W^l(u), \qquad (13.49)$$

where $W^l(u)$ is the total potential energy of the altered body corresponding the same displacement field u. Note that this displacement field is admissible for the altered body, since on the boundary Γ_u it assumes the same zero values as before. If we apply the principle of minimum potential energy for the body [Tim87, Was82, GE94] we can conclude that

$$W^l(u) \geq W^l(u^l). \qquad (13.50)$$

Using Clapeyron's theorem, we can assert that, at equilibrium, the following equalities hold for the original and altered bodies:

$$W(u) = -\frac{1}{2} \langle q, u \rangle = -J, \tag{13.51}$$

$$W^l(u^l) = -\frac{1}{2} \langle q, u^l \rangle = -J^l. \tag{13.52}$$

Using (13.51), (13.52) with (13.49), (13.50), we derive

$$J = \frac{1}{2} \langle q, u \rangle \le \frac{1}{2} \langle q, u^l \rangle = J^l \tag{13.53}$$

and, consequently, the original problem with incomplete information (13.44) is transformed to the conventional optimization problem

$$J_* = \min_{\Gamma \subset H_\Gamma} \frac{1}{2} \langle q, u^l \rangle. \tag{13.54}$$

Part III
Probabilistic and Mixed
Probabilistic – Guaranteed Approaches

Chapter 14
Some Basic Notions of Probability Theory

14.1 Random Variables

A sample space S associated with an experiment is a set of elements such that any outcome of the experiment corresponds to a unique element of the set. An element A is a subset of a sample space S. An element in a sample space is called an event.

An event is called *certain* if it always occurs for a given set of conditions; it is said to be *impossible* if it is known that it never occurs under these conditions. An event which may or may not occur under a given set of conditions is called a *random* event. The objective mathematical estimation of realizing a random event is its *probability*. If an experiment can occur in n mutually exclusive and equally likely ways, and if exactly n_A of these ways correspond to an event A then the ratio

$$v(A) = \frac{n_A}{n} \tag{14.1}$$

is called the *frequency* of the event A. For a sufficiently large number of experiments, n, the value of the frequency $v(A)$ of the event A, may be taken as an approximate measure of the probability $P(A)$, i.e.

$$P(A) \approx v(A) = \frac{n_A}{n}. \tag{14.2}$$

It is evident that

$$0 \leq v(A) \leq 1, \tag{14.3}$$

$$0 \leq P(A) \leq 1, \tag{14.4}$$

where the lower bound is attained for the impossible event, and the upper for a certain event. If S is the sample space, then

$$P(S) = 1. \tag{14.5}$$

N.V. Banichuk and P.J. Neittaanmäki, *Structural Optimization with Uncertainties*, Solid Mechanics and Its Applications 162, DOI 10.1007/978-90-481-2518-0_1, © Springer Science+Business Media B.V. 2010

If $A_1, A_2, \ldots, A_n$ are mutually exclusive events, then

$$P(A_1 \cup A_2 \cup \ldots A_{n-1} \cup A_n) = P(A_1) + P(A_2) + \ldots + P(A_n). \qquad (14.6)$$

If the events $A_1, A_2, \ldots, A_n$ are independent, i.e., if the occurrence of one is independent of the occurrence of the other, then

$$P(A_1 \cap A_2 \cap \ldots A_{n-1} \cap A_n) = P(A_1) P(A_2) \ldots P(A_n). \qquad (14.7)$$

A function whose domain is a sample space S and whose range is some set of real numbers is called a random variable. This random variable is called discrete if it assumes only a finite or denumerable number of values. It is called continuous if it assumes a continuum of values. Suppose that a random variable ξ takes on various values x, within the range $-\infty \leq x \leq \infty$. The probabilistic properties of the random variable ξ may be characterized by using a distribution function $F(x)$, which is the probability of observing the value $\xi < x$ and is written as

$$F(x) = P(\xi < x). \qquad (14.8)$$

Note that the function $F(x)$ is a nondecreasing function and

$$F(-\infty) = 0, \ F(\infty) = 1. \qquad (14.9)$$

The probability that ξ is between x_1 and x_2 is

$$P(x_1 < \xi < x_2) = P(\xi \leq x_2) - P(\xi \leq x_1) = F(x_2) - F(x_1) \qquad (14.10)$$

and

$$F(x_2) \geq F(x_1) \text{ for } x_2 > x_1. \qquad (14.11)$$

Suppose that the random variable is continuous and has a continuous and differentiable probability distribution function, then the derivative

$$f(x) = \frac{dF(x)}{dx} \qquad (14.12)$$

is called the *probability density*. We have $f(x) \geq 0$ and

$$P(\xi < x) = F(x) = \int\limits_{-\infty}^{x} f(x)dx, \qquad (14.13)$$

$$P(x_1 < \xi < x_2) = \int\limits_{x_1}^{x_2} f(x)dx. \qquad (14.14)$$

The probability that ξ is the whole range $-\infty \leq \xi \leq \infty$ is evidently unity, and, consequently, we have the following normalization condition:

$$\int_{-\infty}^{\infty} f(x)dx = 1.$$ (14.15)

Let us introduce some characteristics of the random variable ξ. The *arithmetic mean* $\widehat{\xi}$ is given by the value

$$\widehat{\xi} = E(\xi) = \int_{-\infty}^{\infty} xf(x)dx.$$ (14.16)

Called the mathematical expectation of the random variable ξ. The *second-order moment* of the random variable ξ

$$\widehat{\xi^2} = \int_{-\infty}^{\infty} x^2 f(x)dx$$ (14.17)

is its mean square. The mean-square deviation of the variable ξ from its mean value is called the variance or dispersion and is determined as

$$D(\xi) = E\left(\left(\xi - \widehat{\xi}\right)^2\right) = \int_{-\infty}^{\infty} \left(x - \widehat{\xi}\right)^2 f(x)dx.$$ (14.18)

Note that the variance and *standard deviation*

$$\sigma = \sqrt{D(\xi)}$$ (14.19)

are a measure of the scatter of the random variable around its mean value.

Let $g(\xi)$ be an arbitrary function of the random variable ξ. The *expected value* (or *expectation*) of $g(\xi)$, denoted by $E[g(\xi)]$, is defined by

$$E[g(\xi)] = \int_{-\infty}^{\infty} g(t)f(t)dt.$$ (14.20)

Note that the expected values of the random variables ξ and μ are characterized by the following equations:

$$E(c_1\xi + c_2\eta) = c_1E(\xi) + c_2E(\eta),$$ (14.21)

$$E(\xi\eta) = E(\xi)E(\eta),$$ (14.22)

where c_1, c_2 are arbitrary constants. It is supposed that the random variables ξ and η in (14.22) are statically independent.

Note also the properties of the standard deviation σ:

$$\sigma^2(c\xi) = c^2\sigma^2(\xi), \tag{14.23}$$

$$\sigma^2(c + \xi) = \sigma^2(\xi), \tag{14.24}$$

$$\sigma^2(c_1\xi + c_2) = c_1^2\sigma^2(\xi). \tag{14.25}$$

14.2 Some Continuous Distributions

If the random variable ξ has a density function of the form

$$f(x) = \begin{cases} 0, & x < x_1, \\ \frac{1}{x_2 - x_1}, & x_1 < x < x_2, \\ 0, & x > x_2, \end{cases} \tag{14.26}$$

then the variable ξ is said to possess a *uniform* distribution. It is written as

$$F(x) = \begin{cases} 0, & x < x_1, \\ \frac{x - x_1}{x_2 - x_1}, & x_1 < x < x_2, \\ 1, & x > x_2, \end{cases} \tag{14.27}$$

where x_1, x_2 $(x_1 < x_2)$ are given constants. We have the following properties for the uniform distribution:

$$\widehat{\xi} = \mathrm{E}(\xi) = \frac{x_1 + x_2}{2},$$

$$D(\xi) = \frac{(x_2 - x_1)^2}{12}, \quad \sigma = \sqrt{\frac{(x_2 - x_1)^2}{12}}. \tag{14.28}$$

If the random variable ξ has a density function of the form

$$f(x) = \frac{1}{\sigma\sqrt{2\pi}} \exp\left(-\frac{\left(x - \widehat{\xi}\right)^2}{2\sigma^2}\right), \quad -\infty < x < \infty, \tag{14.29}$$

then the variable ξ is said to possess a *normal (Gaussian)* distribution. It is easy to check that

$$\int_{-\infty}^{\infty} f(x)dx = \frac{1}{2\pi} \int_{-\infty}^{\infty} e^{-t^2/2} dt = 1, \quad t = \frac{x - \widehat{\xi}}{\sigma}$$

and to estimate the mathematical expectation and the variance

$$E(\xi) = \int_{-\infty}^{\infty} xf(x)dx = \widehat{\xi},$$

$$D(\xi) = \int_{-\infty}^{\infty} \left(x - \widehat{\xi}\right)^2 f(x)dx = \sigma^2. \tag{14.30}$$

The probability density of the random variable ξ, described by the normal distribution, is shown in Fig. 14.1 for different values of the standard deviation σ.

If $\widehat{\xi} = 0$, then positive and negative values of the random variable ξ are equally probable and the distribution is called *symmetric*.

The distribution function, corresponding to the probability density (14.29), is

$$F(x) = \frac{1}{\sigma\sqrt{2\pi}} \int_{-\infty}^{\infty} e^{-\frac{\left(x-\widehat{\xi}\right)^2}{2\sigma^2}} dx \tag{14.31}$$

This function can be also represented in the form

$$F(x) = \frac{1}{2} + \Phi\left(\frac{x - \widehat{\xi}}{\sigma}\right) \tag{14.32}$$

where $\Phi(u)$ is *Gauss' error integral*, defined as

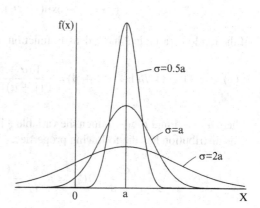

Fig. 14.1 Probability density for a normal distribution

$$\Phi(u) = \frac{1}{\sqrt{2\pi}} \int_0^u e^{-t^2/2} dt. \tag{14.33}$$

The probability that the normally distributed random variable ξ takes on a value within the interval $x_1 \leq \xi \leq x_2$ is determined with the help of the following formula:

$$P(x_1 \leq \xi \leq x_2) = \Phi\left(\frac{x_2 - \widehat{\xi}}{\sigma}\right) - \Phi\left(\frac{x_1 - \widehat{\xi}}{\sigma}\right) \tag{14.34}$$

If the random variable ξ has a density function of the form

$$f(x) = \frac{1}{\Gamma(1+\alpha)\,\beta^{1+\alpha}} x^\alpha e^{-x/\beta}, \ 0 < x < \infty \tag{14.35}$$

with $\alpha > -1$ and $\beta > 0$, then the variable ξ is said to possess a *gamma* distribution. For a gamma distribution it is known that

$$E(\xi) = \beta(1+\alpha),$$
$$D(\xi) = \beta^2(1+\alpha), \ \sigma = \beta\sqrt{1+\alpha}, \tag{14.36}$$
$$F(x) = \frac{\Gamma(1+\alpha, x/\beta)}{\Gamma(1+\alpha)}, \ 0 < x < \infty.$$

If the random variable has a density function of the form

$$f(x) = \frac{1}{\theta} e^{-x/\theta}, \ 0 < x < \infty, \tag{14.37}$$

where $\theta > 0$, then the variable ξ is said to possess an *exponential* distribution. We have the following properties of an exponential distribution:

$$E(\xi) = \theta, \ D(\xi) = \theta^2, \ \sigma = \theta,$$
$$F(x) = 1 - \exp(-x/\theta), \ 0 < x < \infty. \tag{14.38}$$

If the random variable ξ has a density function of the form

$$f(x) = x^\alpha(1-x)^\beta B(1+\alpha, 1+\beta) = \frac{\Gamma(\alpha+\beta+2)}{\Gamma(1+\alpha)\Gamma(1+\beta)} x^\alpha(1-x)^\beta, \ 0 < x < 1, \tag{14.39}$$

where $\alpha > -1$ and $\beta > -1$, then the variable ξ is said to possess a *beta* distribution. This distribution has the following properties:

$$E(\xi) = \frac{1+\alpha}{2+\alpha+\beta}, \ D(\xi) = \sigma^2 \frac{(1+\alpha)(1+\beta)}{(2+\alpha+\beta)^2(3+\alpha+\beta)}. \tag{14.40}$$

If the random variable ξ has a density function of the form

$$f(x) = \alpha \beta x^{\alpha-1} \exp\left(-\beta x^{\alpha}\right), \ 0 < x < \infty \tag{14.41}$$

then the variable ξ is said to possess a *Weibull* distribution. This distribution has the following properties:

$$F(x) = 1 - \exp(-\beta x^{\alpha}), \ 0 < x < \infty,$$

$$E(\xi) = \Gamma\left(1 + \frac{1}{\alpha}\right)\beta^{-1/\alpha}, \tag{14.42}$$

$$D(\xi) = \left[\Gamma(1 + \frac{2}{\alpha}) - \Gamma^2\left(1 + \frac{1}{\alpha}\right)\right]\beta^{-2/\alpha}.$$

If the random variable ξ has a density function of the form

$$f(x) = \frac{\alpha^{\lambda}}{\Gamma(\lambda)} x^{\lambda-1} e^{-\beta x}, \ 0 < x < \infty \tag{14.43}$$

then the variable ξ is said to possess a *chi-square* $\left(\chi^2\right)$ *distribution*. This distribution has the following properties:

$$F(x) = \frac{\Gamma(\lambda, \alpha x)}{\Gamma(\lambda)}, \ 0 < x < \infty,$$

$$E(\xi) = \frac{\lambda}{\alpha}, \ D(\xi) = \sigma^2 = \frac{\lambda}{\alpha^2}. \tag{14.44}$$

If the random variable ξ has a density function of the form

$$f(x) = \frac{x}{\gamma^2} \exp\left(-\frac{x^2}{2\gamma^2}\right), \ 0 < x < \infty \tag{14.45}$$

then the variable ξ is said to possess a *Rayleigh* distribution. This distribution has the following properties:

$$F(x) = 1 - e^{-x^2/2\gamma^2}, \ 0 < x < \infty,$$

$$E(\xi) = \gamma\sqrt{\pi/2}, \ D(\xi) = \gamma^2(4 - \pi)/2. \tag{14.46}$$

14.3 Functions of Random Variables

Consider a function

$$\eta = g(\xi) \tag{14.47}$$

of one random argument ξ. By definition

$$F(y) = P(\eta < y) = P(g(\xi) < y). \tag{14.48}$$

From Eq. (14.48), it follows that

$$F(y) = \int\limits_{g(x)<y} f(x)dx \tag{14.49}$$

where $f(x)$ is the probability density function of random variable ξ and $F(y)$ is the distribution function of random variable η. The symbol below the integral in (14.49) indicates that the integration is performed for all intervals of the real axis on which $g(x) < y$.

To find the probability density function of random variable η let us consider the interval

$$x < \xi < x + dx \tag{14.50}$$

and corresponding interval

$$y < \eta < y + dy \tag{14.51}$$

obtained with the help of the function (14.47). The probabilities of satisfying the inequalities (14.50) and (14.51) are equal. Taking into account that these probabilities correspond to the dashed areas (in Fig. 14.2) $f(x)dx$ and $f(y)dy$, we will have

$$f(x)dx = f(y)dy. \tag{14.52}$$

From here we determine the simplest formula of probabilities transformation

$$f(y) = \frac{f(x)}{\left|\frac{dy}{dx}\right|}. \tag{14.53}$$

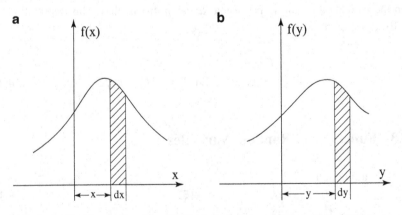

Fig. 14.2 Probability density functions $f(x)$ and $f(y)$ of the random variables ξ and η

We suppose here that the function $g(\xi)$ is monotone and consequently the inverse function $\xi = \zeta(\eta)$ is single-valued. Using (14.49) and (14.53), we obtain

$$F(y) = \int\limits_{-\infty}^{y} f(\zeta(y)) \left| \frac{d\zeta(y)}{dy} \right| dy, \tag{14.54}$$

$$f(y) = f(\zeta(y)) \left| \frac{d\zeta(y)}{dy} \right|. \tag{14.55}$$

Let us consider a two-dimensional case, when we have two random variables ξ and ζ, which are not generally independent. The joint two-dimensional distribution function of these random variables is the function

$$P(\xi < x, \zeta < z) = F(x, z). \tag{14.56}$$

The joint distribution function $F(x, z)$ and the joint probability density $f(x, z)$ are related by means of the relationship

$$f(x, z) = \frac{\partial^2 F(x, z)}{\partial x \partial z}, \tag{14.57}$$

$$F(x, z) = \int\limits_{-\infty}^{x} \int\limits_{-\infty}^{z} f(x, z) dx dz. \tag{14.58}$$

The probability of obtaining a random variable (ξ, ζ) in a two-dimensional domain S is defined by the formula

$$P = \iint\limits_{S} f(x, z) dx dz. \tag{14.59}$$

Hence, the normalization condition becomes

$$\int\limits_{-\infty}^{\infty} \int\limits_{-\infty}^{\infty} f(x, z) dx dz = 1. \tag{14.60}$$

The joint probability density $f(x, z)$ of statistically independent random variables ξ and ζ is the product of the partial probability densities $f_\xi(x)$, $f_\zeta(z)$, i.e.

$$f(x, z) = f_\xi(x) f_\zeta(z), \tag{14.61}$$

where

$$f_\xi(x) = \int\limits_{-\infty}^{\infty} f(x,z)dz, \quad f_\zeta(z) = \int\limits_{-\infty}^{\infty} f(x,z)dx. \tag{14.62}$$

The corresponding distribution function is given by

$$F(x,z) = F_\xi(x)F_\zeta(z), \tag{14.63}$$

where

$$F_\xi(x) = \int\limits_{-\infty}^{x} \int\limits_{-\infty}^{\infty} f(x,z)dxdz,$$

$$F_\zeta(z) = \int\limits_{-\infty}^{\infty} \int\limits_{-\infty}^{z} f(x,z)dxdz. \tag{14.64}$$

If the random variables ξ, ζ are statistically independent random variables and are distributed according to a normal law

$$f_\xi(x) = \frac{1}{\sqrt{2\pi}\sigma_\xi} \exp\left[-\frac{\left(x-\widehat{\xi}\right)^2}{2\sigma_\xi^2}\right],$$

$$f_\zeta(z) = \frac{1}{\sqrt{2\pi}\sigma_\zeta} \exp\left[-\frac{\left(z-\widehat{\zeta}\right)^2}{2\sigma_\zeta^2}\right] \tag{14.65}$$

with given mean values $\widehat{\xi}, \widehat{\zeta}$ and standard deviations σ_ξ, σ_ζ, then

$$f(x,z) = \frac{1}{2\pi\sigma_\xi\sigma_\zeta} \exp\left\{-\frac{1}{2}\left[\frac{\left(x-\widehat{\xi}\right)^2}{\sigma_\xi^2} + \frac{\left(z-\widehat{\zeta}\right)^2}{\sigma_\zeta^2}\right]\right\} \tag{14.66}$$

The probability density $f_\eta(y)$ for a sum $\eta = \xi + \zeta$ of random variables ξ and ζ is given by

$$f_\eta(y) = \frac{1}{2\pi\sigma_\eta^2} \exp\left[-\frac{(y-\widehat{\eta})^2}{2\sigma_\eta^2}\right], \tag{14.67}$$

$$\widehat{\eta} = \widehat{\xi} + \widehat{\zeta}, \quad \sigma_\eta^2 = \sigma_\xi^2 + \sigma_\zeta^2. \tag{14.68}$$

The joint distribution of the random variables ξ, ζ may be described by their mean values $\widehat{\xi}$, $\widehat{\zeta}$

$$\widehat{\xi} = \mathrm{E}(\xi) = \int\limits_{-\infty}^{\infty} x f_\xi(x)dx,$$

$$\widehat{\zeta} = \mathrm{E}(\zeta) = \int\limits_{-\infty}^{\infty} z f_\zeta(z)dz \qquad (14.69)$$

and second-order moments $\overline{\xi\xi}$, $\overline{\zeta\zeta}$, $\overline{\xi\zeta}$

$$\overline{\xi\xi} = \int\limits_{-\infty}^{\infty} x^2 f_\xi(x)dx, \quad \overline{\zeta\zeta} = \int\limits_{-\infty}^{\infty} z^2 f_\zeta(z)dz,$$

$$\overline{\xi\zeta} = \int\limits_{-\infty}^{\infty}\int\limits_{-\infty}^{\infty} xz f(x,z)dxdz. \qquad (14.70)$$

The corresponding central moments $K_{\xi\xi}$, $K_{\zeta\zeta}$, $K_{\xi\zeta}$, given by the formulas

$$K_{\xi\xi} = \mathrm{E}\left(\left(\xi - \widehat{\xi}\right)^2\right) = \int\limits_{-\infty}^{\infty} \left(x - \widehat{\xi}\right)^2 f_\xi(x)dx \qquad (14.71)$$

$$K_{\zeta\zeta} = \mathrm{E}\left(\left(\zeta - \widehat{\zeta}\right)^2\right) = \int\limits_{-\infty}^{\infty} \left(z - \widehat{\zeta}\right)^2 f_\zeta(z)dz \qquad (14.72)$$

$$K_{\xi\zeta} = \mathrm{E}\left(\left(\xi - \widehat{\xi}\right)\left(\zeta - \widehat{\zeta}\right)\right) = \int\limits_{-\infty}^{\infty}\int\limits_{-\infty}^{\infty} \left(x - \widehat{\xi}\right)\left(z - \widehat{\zeta}\right) f(x,z)dxdz \qquad (14.73)$$

are called the *correlation moments* and the matrix K, composed of these moments is called the *correlation matrix*. The correlation matrix K is symmetric, i.e.

$$K_{\xi\zeta} = K_{\zeta\xi} \qquad (14.74)$$

and its diagonal elements are the variances of the corresponding random variables.

The mixed moment $K_{\xi\zeta}$ characterizes the statistical connection (correlation) between the random variables ξ and ζ. For statistically independent random variables, when $f(x,z) = f_\xi(x)f_\zeta(z)$, we will have

$$K_{\xi\zeta} = \mathrm{E}\left(\left(\xi - \widehat{\xi}\right)\left(\zeta - \widehat{\zeta}\right)\right) = \int\limits_{-\infty}^{\infty} \int\limits_{-\infty}^{\infty} \left(x - \widehat{\xi}\right)\left(z - \widehat{\zeta}\right) f(x,z)dxdz$$

$$= \left\{\int\limits_{-\infty}^{\infty} xf_{\xi}(x)dx - \widehat{\xi}\right\}\left\{\int\limits_{-\infty}^{\infty} zf_{\zeta}(z)dz - \widehat{\zeta}\right\} = 0 \qquad (14.75)$$

and, consequently there is no correlation between ξ and ζ.

Chapter 15
Probabilistic Approaches for Incomplete Information

Most structural optimization problems that have been investigated use approaches designed for complete data. Nevertheless, some parametric probabilistic problems of optimal structural design with limiting state and buckling constraints have been formulated and investigated in [AB73, ABC84, AC79, Bra84, DFH77, ETB84, AGRZ84, EKS86, Mos77, Mar95, BV07]. Nondeterministic approaches have been used for some optimal design problems when the external conditions (external forces) were taken as a random variables. Fewer probabilistic studies have been devoted to the important class of problems of quasi-brittle elastic body optimization on the basis of modern fracture mechanics criteria [ABC84, BRS99, BRS03a, BRS03b, LM96]. In accordance with fracture mechanics representations it is necessary to consider all possibilities for the appearance of cracks, and to take into account that the crack position, size and mode (opening cracks, shear cracks, etc.) are unknown beforehand. Therefore, the formulation and solution of these optimization problems requires the application of methods which can take into account incomplete information. Minimax methods of game theory (guaranteed approach), probabilistic methods, etc., can be used for this purpose.

This chapter deals with the probabilistic approach to optimal design problems for quasi-brittle structural elements. Special attention is devoted to different problem formulations and analytical solution methods. Using these methods, we find optimal designs of beams with surface and internal cracks. It is possible to apply the guaranteed approaches of game theory, as it was done earlier in Chapters 1–13, to optimize the structures for the worst case scenario. But it is necessary to consider all possible crack appearances, as required by the guaranteed approaches [CK78, BRS99, BRS03a, BRS03b, BH75, CM78, Bol61, Bol69]. Besides the complexity of these approaches, it is also important from a practical point of view to note that the optimal design for the worst case scenario entails larger structural cost (weight or other functionals) than that obtained from structural optimization based on probabilistic approaches. This chapter develops such probabilistic approaches which produce more optimistic results. This chapter provides methods for finding the optimal design for many cases when the probability density distributions for random variables are known. The presentation follows research results of [BRS99].

N.V. Banichuk and P.J. Neittaanmäki, *Structural Optimization with Uncertainties*, Solid Mechanics and Its Applications 162, DOI 10.1007/978-90-481-2518-0_1, © Springer Science+Business Media B.V. 2010

15.1 Probabilistic Problems with Fracture Mechanics Constraints

Consider a deformed structural element occupying a two-dimensional domain and loaded by given external forces. The stressed state is described by the equations of elasticity. The body consists of brittle or quasi-brittle material, and contains a crack of length l. The crack is supposed to be rectilinear and small with respect to the characteristic size L of the body ($l \ll L$). The crack center coordinates x_c, y_c, the angle of inclination of the crack with respect to the axis x of the global coordinate system (x, y), the length of the crack, or some of these parameters, are unknown beforehand. For design purposes we must consider a variety of crack locations and sizes and solve corresponding elasticity equations. If we can solve these equations we can estimate the stress intensity factors K_1, K_2 occurring in asymptotic representations [Hel84, Hut79, GE94, KP85, Mor84, Par92, BPB$^+$00]

$$\sigma_n = \frac{K_1}{\sqrt{2\pi r}} + O(1), \quad \sigma_{nt} = \frac{K_2}{\sqrt{2\pi r}} + O(1) \tag{15.1}$$

for the components of the stress tensor at the point that lies on the crack axis nearest to the crack tip (r is the distance of the point t of the axis from the crack tip). Here the t-axis of the local orthogonal coordinate system (t, n) is parallel to the crack, while the origin is placed at the middle point of the crack. The condition for the crack not to propagate, is known from brittle and quasi-brittle fracture mechanics: it has the form

$$K_1 \le K_{1C} \tag{15.2}$$

for an opening crack, the form

$$K_2 \le K_{2C} \tag{15.3}$$

for a shear crack, the form

$$\frac{K_1^2 + K_2^2}{E} \le G_C \tag{15.4}$$

for the complex case, where K_{iC} ($i = 1, 2$) and G_C are given brittle strength constants.

In this chapter we will consider one crack parameter as a random variable ξ with known probability density function $f(\xi)$, while the other crack parameters will be considered as given or will be found with the help of the guaranteed approach of game theory (design for the worst case). Let us suppose that the shape of the body is described by a function h of space coordinates that is taken as the unknown design variable. Crack mode $i = 1$ is an opening crack, $i = 2$ is a shear crack. If we introduce the small positive parameter ε and write the limiting stress intensity factor as $K_{i\varepsilon} = K_{1C} - \varepsilon_i, \varepsilon_i \ge 0$ [Bol69, Str47] then we can write the fracture mechanics constraint in the following form:

$$K_i(\xi, h) = K_{i\varepsilon} \tag{15.5}$$

In general terms the stress intensity factor $K_i(\xi, h)$ depends implicitly (through the dependence on crack parameters and the solution of the elasticity equations) on a random variable ξ, but for particular cases, which will be discussed in the next part of the chapter, we can simplify the dependence of K_i on ξ and express K_i in some explicit functional form. The probability density distribution $f(\xi)$ for the variable ξ permits us, at least in principle, to determine the moments of the random quantity K_i and, in particular, their mathematical expectation and dispersion (variance). This in turn allows us to control the probability of violating the assigned strength constraint (15.5) and to study the problem of optimal design, consisting in finding a function h that satisfies the system of inequalities

$$\widehat{K}_i = E(K_i) = \int_{-\infty}^{\infty} K_i(\xi, h) f(\xi) d\xi \leq K_{i\varepsilon} \qquad (15.6)$$

$$D(K_i) = E\left(\left(K_i - \widehat{K}_i\right)^2\right) = \int_{-\infty}^{\infty} \left(K(\xi, h) - \widehat{K}\right)^2 f(\xi) d\xi \leq \delta \qquad (15.7)$$

and minimizes the functional $J(h)$ (volume or weight of the structural element)

$$J = J(h) \to \min_h \qquad (15.8)$$

while taking care of the equations of the theory of elasticity. Here E and D denote mathematical expectation and dispersion, respectively, for a random variable ξ, and δ denotes a sufficiently small number which is supposed to be given. The inequalities (15.6) and (15.7) imply that some conditions must accompany a choice of the design variable h. The brittle fracture mechanics constraint given in (15.5) must be satisfied in an average sense, and the deviation of quantity $K_i - \widehat{K}_i$ must not exceed some small chosen value.

Now consider another approach [BRS99] for the formulation of the structural optimization problem using the theory of probability. Let us suppose additionally that $K_i(\xi, h)$ is a monotonic function of ξ, and the equation

$$K_i(\xi, h) = K_0$$

has the single solution

$$\xi_0 = \xi_0(h, K_0)$$

for fixed h and K_0. Then the probability P of satisfying the inequality

$$K_i \leq K_0$$

can be estimated by the following relation:

$$P\{K_i \leq K_0\} = P\{\xi \leq \xi_0\} = \int_0^{\xi_0} f(\xi) d\xi. \qquad (15.9)$$

The probabilistic expression for the condition for the crack not to propagate can be written in the form

$$P\{K_i \leq K_{i\varepsilon}\} = \int_0^{\xi_\varepsilon} f(\xi)d\xi \geq 1 - \nu \qquad (15.10)$$

where $K_{i\varepsilon} = K_{iC} - \varepsilon$ and $\nu > 0$ is a small given positive value. The variable ξ_ε is the solution of the equation

$$K_i(\xi, h) = K_{i\varepsilon} \qquad (15.11)$$

An optimization problem based on fracture mechanics and probability can be formulated as a problem of minimization of the cost functional (15.8) under the constraint (15.10) and other stiffness constraints, taking into account the elasticity equations for the structural elements.

15.2 Beams with Random Crack Length

Consider the optimal design of a beam subjected to the action of transverse loads taking into account possible cracks at the beam surfaces. We assume that the beam of length L lies along the x-axis ($0 \leq x \leq L$) and has a rectangular cross-section with height $h = h(x)$ and constant width b. The bending moment $M(x)$ and shear force $Q(x)$ acting on the cross-section of the beam are independent of the elastic properties and the shape of the cross-section and are considered as known for the interval $0 \leq x \leq L$. The maximum tensile stress is attained on the elongated beam surface for any cross-section, that is $|\zeta| = h(x)/2$, where the coordinate ζ measures the distance from the center of the cross-section and varies in the interval

$$-h(x)/2 \leq \zeta \leq h(x)/2.$$

The maximum stress is

$$\sigma_x = 6|M(x)|/bh^2(x).$$

The stress intensity factor K_1 for a surface crack at any cross-section can be evaluated using the formula [KP85]

$$K_1 = 1.12\sigma_x \sqrt{\pi \xi} = \varphi(x)\sqrt{\xi},$$
$$\varphi(x) = \frac{k_1|M(x)|}{bh^2(x)}, \qquad (15.12)$$

where $k_1 = 6.72\sqrt{\pi}$, and the length of the surface crack ξ is a random variable characterized by the known density function $f(\xi)$. For the mean value of K_1 (the mathematical expectation $E(K_1)$) and the variance $D(K_1)$ we will have

$$E(K_1) = \int_0^\infty K_1 f(\xi) d\xi = \varphi(x) E\left(\sqrt{\xi}\right), \tag{15.13}$$

$$D(K_1) = \int_0^\infty \left((K_1 - E(K_1))^2\right) f(\xi) d\xi = \varphi^2(x) D\left(\sqrt{\xi}\right), \quad 0 \le x \le L. \tag{15.14}$$

Using the expressions (15.12)–(15.14), we can present the probabilistic fracture mechanics constraints (15.6), (15.7) in the form

$$h(x) \ge \Psi_E \equiv \left[\frac{k_1 |M(x)|}{b K_{1\varepsilon}} E\left(\sqrt{\xi}\right)\right]^{1/2}, \tag{15.15}$$

$$h(x) \ge \Psi_D \equiv \left[\frac{k_1 |M(x)|}{b}\right]^{1/2} \left[\frac{E\left(\sqrt{\xi}\right)}{\delta}\right]^{1/4}. \tag{15.16}$$

The optimization problem consists in finding an admissible thickness distribution such that the cost functional is minimized, while satisfying the transformed strength constraints (15.15), (15.16). The solution of the problem is written in explicit form

$$h(x) = \begin{cases} \Psi_E(x) & \text{if } \frac{E(\sqrt{\xi})}{K_{1\varepsilon}} \ge \left[\frac{D(\sqrt{\xi})}{\delta}\right]^{1/2}, \\ \Psi_D(x) & \text{if } \frac{E(\sqrt{\xi})}{K_{1\varepsilon}} < \left[\frac{D(\sqrt{\xi})}{\delta}\right]^{1/2} \end{cases} \tag{15.17}$$

or in more compact form

$$h(x) = \max \{\Psi_E(x), \Psi_D(x)\}. \tag{15.18}$$

Let us take into account the geometrical constraint on the thickness distribution $h \ge h_{\min}$ ($h_{\min}$ is a given constant). The optimal solution can be written as

$$h(x) = \max \{h_{\min}, \Psi_E(x), \Psi_D(x)\}. \tag{15.19}$$

We now consider the optimal design of brittle and quasi-brittle beams with low limiting value of the shear stress intensity factor K_{2C}. Shear stresses are maximal at the neutral axis of the beam. We use the expression

$$\sigma_{xy} = \frac{3 |Q(x)|}{2bh(x)}$$

to obtain a direct functional dependence of maximal shear stresses on the variables b, h, and $Q(x)$. Suppose that the internal shear crack is located on the x-axis, and its length is much smaller that the minimum thickness of the beam h_{min}, and than the characteristic size of the domain of shear stress variation for the uncracked beam. In accordance with the formula for the shear stress intensity factor of an isolated crack (mode II) we have

$$K_2 = \sigma_{xy}\sqrt{\pi\xi} = \varphi(x)\sqrt{\xi}, \tag{15.20}$$

$$\varphi(x) = \frac{k_2\,|Q(x)|}{bh(x)}, \tag{15.21}$$

where $k_2 = 3\sqrt{\pi}/2$. Performing the same estimations as in (15.13)–(15.17) and using the notations

$$\Psi_E(x) \equiv k_2\frac{|Q(x)|\,\mathrm{E}\left(\sqrt{\xi}\right)}{bK_{2\varepsilon}}, \tag{15.22}$$

$$\Psi_D(x) \equiv k_2\frac{|Q(x)|}{b}\left[\frac{\mathrm{E}\left(\sqrt{\xi}\right)}{\delta}\right]^{1/2}, \tag{15.23}$$

we obtain the explicit optimal solution in the form (15.19) which minimizes the optimized functional (volume of the beam) under the fracture mechanics constraint and the requirement $h(x) \geq h_{min}$ ($0 \leq x \leq L$).

15.3 Beams with Randomly Placed Cracks

Consider the optimal design of a beam with rectangular cross-section (width b is the given constant, height $h(x)$ is the unknown design variable, $0 \leq x \leq L$) and let us assume that the material of the beam has a low limiting value of the shear stress intensity factor K_{2C}. The crack lies on the neutral axis and its center coordinate ξ is taken as a random variable with known density function $f(\xi)$. The length l of the crack is known.

In this case

$$K_2 = K_2\,(\xi, h(\xi)) = k_2\frac{|Q(\xi)|}{bh(\xi)}\sqrt{l},$$

where the shear force $Q(\xi)$ is given for $0 \leq x \leq L$. For mean value and variance of K we will have

$$E(K_2) = \int_0^L K_2 f(\xi) d\xi \equiv J_1,$$

$$D(K_2) = \int_0^L \left(K_2 - \int_0^L K_2 f(\xi) d\xi \right)^2 f(x) dx = \qquad (15.24)$$

$$= \int_0^L K_2^2 f(x) dx - \left(\int_0^L K_2 f(x) dx \right)^2 \equiv J_2 - J_1^2,$$

where J_1, J_2 are homogeneous functionals of h

$$J_1(h) = \int_0^L \frac{\psi}{h} f(\xi) d\xi,$$

$$J_2(h) = \int_0^L \frac{\psi^2}{h^2} f(\xi) d\xi, \qquad (15.25)$$

$$\psi = \psi(\xi) = k_2 \frac{|Q(\xi)|}{b} \sqrt{l}.$$

Fracture mechanics constraints are written as inequalities for these integral functionals

$$J_1(h) \le K_{2\varepsilon}, \qquad (15.26)$$

$$J_2(h) - J_1^2(h) \le \delta. \qquad (15.27)$$

Here $K_{2\varepsilon} = K_{2C} - \varepsilon$ and $\varepsilon > 0$ is a small parameter. We must minimize the functional with respect to design variable h under the constraints (15.26), (15.27). For the sake of convenience, in the derivation of necessary optimality conditions, we shall replace the inequality constraints (15.26), (15.27) with equality constraints by introducing auxiliary unknowns (or slack variables) μ_1^2, μ_2^2:

$$J_1 - K_{2\varepsilon} + \mu_1^2 = 0, \qquad (15.28)$$

$$J_2 - J_1^2 - \delta + \mu_2^2 = 0. \qquad (15.29)$$

If strict inequalities are enforced in (15.26), (15.27) for the optimal solution, then $\mu_1 \ne 0$, $\mu_2 \ne 0$. However, if these constraints are "active" and the equal sign occurs in (15.26), (15.27), then

$$\mu_1 = 0, \quad \mu_2 = 0.$$

Let us take into account (15.28), (15.29) and extend our optimized cost functional, making use of Lagrangian multipliers λ_1, λ_2. By varying h and μ_1, μ_2, we obtain the formula of design sensitivity analysis for the volume (weight) functional

$$\delta J = \int_0^L \left[b + (2\lambda_2 J_1 - \lambda_1) \frac{\psi}{h^2} f - 2\lambda_2 \frac{\psi^2}{h^3} f \right] \delta h dx + 2\lambda_1 \mu_1 \delta \mu_1 + 2\lambda_2 \mu_2 \delta \mu_2.$$

$$(15.30)$$

A necessary optimality condition for minimizing the functional J assumes the form $\delta J = 0$. As a consequence, we can derive necessary optimality conditions

$$h^3 + (2\lambda_2 J_1 - \lambda_1) \frac{\psi}{b} f h - 2\lambda_2 \frac{\psi^2}{b} f = 0, \qquad (15.31)$$

$$\lambda_1 \mu_1 = 0, \quad \lambda_2 \mu_2 = 0. \qquad (15.32)$$

We observe that the second coefficient appearing in the necessary optimality conditions for an extremum (15.31) is derived for the value J_1 that corresponds to the extremal value (i.e. to the solution) of the optimization problem. The solution of Eq. 15.31 depends on unknown parameters λ_1, λ_2 and β ($\beta = J_1$), i.e.

$$h = h(x, \lambda_1, \lambda_2, \beta).$$

Determination of the parameter β is achieved with the help of the condition

$$\int_0^L \frac{\psi f dx}{h(x, \lambda_1, \lambda_2, \beta)} = \beta. \qquad (15.33)$$

The quantities λ_1, λ_2 and μ_1, μ_2 can be determined by solving Eqs. (15.28), (15.29), (15.32). The conditions (15.32) imply that for inactive constraints, $\mu_i \neq 0$. The corresponding Lagrangian multipliers λ_i must be equal to 0, and consequently the corresponding constraints can be ignored in all subsequent developments. However, if $\lambda_i \neq 0$, then $\mu_i = 0$ and the constraints indexed by i turns out to be "active". For the sake of brevity we consider only the case in which the constraint in (15.28) is active and the constraint (15.29) is inactive. In this case the solution has the form

$$h_* = \frac{k_2 \alpha_1 \sqrt{l}}{b K_{2\varepsilon}} \sqrt{|Q| f},$$

$$V_* = \frac{k_2 \alpha_1^2 \sqrt{l}}{K_{2\varepsilon}}, \quad \alpha_1 = \int_0^L \sqrt{|Q| f} dx \qquad (15.34)$$

and realizes rigorous equality in the inequality (15.26). Let us substitute the thickness distribution (15.34) into the inequality (15.27) to derive the condition for problem parameters, corresponding to the case considered. We will have

$$\alpha_2 \leq \left(1 + \frac{\delta}{k_{2\varepsilon}^2}\right)\alpha_1^2, \quad \alpha_2 = \int\limits_0^L |Q|\,dx. \tag{15.35}$$

Note that the cases when the constraint (15.26) is inactive, but the constraint (15.27) is active, or both constraints are active can be investigated numerically using a gradient projection algorithm and the basic relation of design sensitivity analysis (15.31), (15.32); or analytically using the explicit expression for the solution of the cubic equation (15.31).

15.4 The Probabilistic Approach with Constraint on the Stress Intensity Factor

First, we shall suppose that the stress intensity factor depends monotonically on a random variable. Let a beam of length L lie along the x-axis ($0 \leq x \leq L$) and have a rectangular cross-section with height $h = h(x)$ (design variable) and given constant width b. Suppose that the beam is loaded by transverse loads, and that a crack of random length ξ can appear at the beam surfaces. The density function for ξ is supposed to be known. The expression (15.12) will be taken for K_1. Taking into account the monotonicity of K_1 with respect to the random variable ξ, we obtain the unique solution ξ_ε of the equation $K = K_{1\varepsilon}$ for every cross-section of the beam

$$\xi_\varepsilon = \frac{b^2 K_{1\varepsilon}^2 h^4(x)}{k_1^2 M^2(x)}. \tag{15.36}$$

For the probabilistic inequality corresponding to brittle fracture, the mechanics criterion can be written in the form

$$P\{K_1 \leq K_{1\varepsilon}\} = P\{\xi \leq \xi_\varepsilon\} = \int\limits_0^{\xi_\varepsilon} f(\xi)d\xi \equiv F(\xi_\varepsilon) \geq 1 - \nu, \tag{15.37}$$

where $\nu > 0$ is a given small positive value, F is the distribution function determined with the help of analytic expression or tabulized data for the known density function f. We must find the thickness distribution such that the volume of the beam is minimized, while the probabilistic strength constraint (15.37) is satisfied. The optimal solution of the problem corresponds to rigorous equality in (15.37), and can be written in the following manner:

$$h(x) = \left[\frac{k_1 |M(x)|}{bK_{1\varepsilon}}\right]^{1/2} (S(1 - \nu))^{1/4}, \tag{15.38}$$

where $S(1 - v)$ is the inverse function with respect to the function F, i.e.

$$S(F(\xi)) = \xi$$

for any ξ.

Now consider the general case of nonmonotonic dependence of the stress intensity factor on a random variable. We investigate the shape optimization problem for a beam with low material constant K_{2C} and suppose that the coordinate ξ of a shear crack center is a random variable with given density function $f(\xi)$. In this case we have the expression (15.20), (15.21) for the stress intensity factor K_2 and can introduce the value

$$\Delta K_2 = K_{1\varepsilon} - K_2 = K_{2\varepsilon} - \frac{k_2 |Q(\xi)| / \sqrt{l}}{bh(\xi)}. \tag{15.39}$$

The length of the crack l is supposed to be given. The inequality $\Delta K_2 > 0$, that guarantees that the crack will not propagate, is written in the form

$$h \geq \frac{k_2 |Q(\xi)| / \sqrt{l}}{bK_{2\varepsilon}} \equiv \chi(x). \tag{15.40}$$

The corresponding probability can be estimated with the help of the integration of the density function $f(\xi)$ over the intervals where $\Delta K_2 \geq 0$ and is written in the following manner:

$$P\{K_2 \leq K_{2\varepsilon}\} = \int_0^\infty g(\xi) f(\xi) d\xi,$$

$$g = \begin{cases} 1 \text{ if } h \geq \chi, \\ 0 \text{ if } h < \chi. \end{cases} \tag{15.41}$$

The optimization problem consists in the determination of the thickness distribution under constraint

$$\int_0^\infty g(\xi) f(\xi) d\xi < 1 - v \tag{15.42}$$

such that $h \geq h_{\min}$ ($h_{\min}$ is a given positive constant) and the cost functional (volume) is minimum. It can be proved that material is removed for the intervals where the density function f is relatively small. For example, if we consider a cantilever beam with monotonically decreasing functions $Q(x)$, $f(x)$ ($0 \leq x \leq L$) shown respectively by dashed lines 1 and 2 in Fig. 15.1, the thickness distribution is described by the solid line.

Fig. 15.1 Optimal
distribution $h(x)$ (*solid line*)

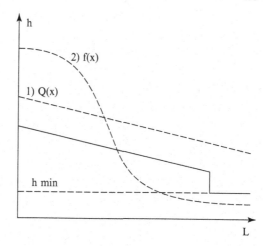

15.5 Conclusions

We have presented some results of structural optimization based on brittle fracture
mechanics, and shown the usefulness of probabilistic approaches and asymptotic
methods. All the chosen structural optimization problems were related to the design
of beams. The optimization method described here can also be applied to optimal design of different one-dimensional structural elements (rods, columns, ties, etc.) and
two-dimensional thin walled structural elements (plates and shells). We have con-
fined our attention to cases in which crack characteristics depend only on a scalar
random variable. It is clear that the moment and direct probabilistic approaches can
be generalized and applied to more difficult cases when the crack parameters are
functions of a random vector. In this case the moment approach will also include
constraints for correlation moments. As for the direct approach, it will use multidi-
mensional analysis of the regions where

$$K_i(\xi_1, \xi_2, \ldots, \xi_m) \le K_{iC}$$

to estimate the probability that cracks will not propagate. The particular case of
independent random variables with given density functions, and correlation characteristics identically zero, provides a meaningful simplification for optimization
problems. We have also focused here on the case in which external loads do not
vary. The important case of cyclic loading will be considered in Chapter 16.

Chapter 16
Optimization Under Longevity Constraint

This chapter deals with probabilistic approaches to optimal design of structures made from quasibrittle material and loaded by cyclic forces [YCC97, PKK97, BIMS05a, BIMS05b, BIM07]. Special attention is devoted to different problem formulations and analytical solution methods. First we present some basic assumptions and relations. Then we formulate the optimal structural design problem based on a probabilistic approach. We must minimize the cost functional (volume of material) under constraints on the number of loading cycles before global fracture and on the probability of nondestructive behavior of the body. The original constraints are transformed to inequalities imposed on the stress in the uncracked body at the crack location. The resulting problem of optimal shape design consists of cost functional minimization under stress constraints, and can be solved by conventional methods. Several examples of structural design problems for statically determinate and indeterminate beams and frames are presented in the chapter. Then we use another probabilistic approach, based on the application of moment inequalities, for optimal structural design under a longevity constraint (constraint on the number of cycles). Here we require that the mathematical expectation (first moment) of the critical number of cycles must be greater than the given number of cycles, and the dispersion (second movement) of the critical number of cycles must be less than a given value. It is shown that this problem can be transformed to that of the structural volume minimization under a system of stress constraints. The presentation follows research results of [BRS03a].

16.1 Basic Assumptions and Relations

Consider an elastic body subjected to quasistatic cyclic loading

$$q = q_0(x)p, \ 0 \leq p_{min} \leq p \leq p_{max},$$

where p is a loading parameter; p_{min}, p_{max} are given values, and $q_0(x)$ is a given amplitude function of the applied load. The body consists of quasi-brittle material

N.V. Banichuk and P.J. Neittaanmäki, *Structural Optimization with Uncertainties*, Solid Mechanics and Its Applications 162, DOI 10.1007/978-90-481-2518-0_1,

and contains a crack of length l. The crack is supposed to be rectilinear and small with respect to the characteristic size L of the body ($l \ll L$).

The initial length l_i of the crack (before loading) is considered as a random variable ξ ($\xi = l_i$), and its position is not also fixed. After application of cyclic loading, the initial crack propagates and its length l increases monotonically. Fatigue crack growth loading can be adequately characterized in the following form (see [Hel84, KP85, Ser00, SL68, Smi91]):

$$\frac{dl}{dn} = C(\Delta K_1)^\alpha, \quad l_i \le l \le l_{cr}, \; 0 \le n \le n_{cr}. \tag{16.1}$$

Here C and α are some material constants ($2 \le \alpha \le 4$), K_1 is the stress intensity factor for the opening crack computed with the help of the formula $K_1 = k\sigma\sqrt{\pi l}$, and the increment ΔK_1 is given by

$$\Delta K_1 = (K_1)_{\max} - (K_1)_{\min} = k\sqrt{\pi l}\sigma_0 (p_{\max} - p_{\min}), \tag{16.2}$$

where k is the configuration correction factor, $\sigma_0 = (\sigma)_{p=1}$, and $(K_1)_{\max}$, $(K_1)_{\min}$ are the maximum and minimum values of the stress intensity factor K_1 in any given cycle, respectively. The ordinary differential equation (16.1) defines quasistatic crack growth, and determines the dependence of the crack length l on the number of cycles n. This equation is valid up to the moment $n = n_{cr}$, where $l = l_{cr}$ and the unstable crack growth (catastrophic fracture of the structure) is attained.

To find l_{cr} we will use the fracture criterion

$$K_1(l_{cr}, \sigma_{\max}) = K_{1C}. \tag{16.3}$$

Here K_{1C} is the fracture toughness of the material, $\sigma_{\max}$ is the maximum stress in the uncracked body at the crack location

$$\sigma_{\max} = p_{\max}(\sigma)_{p=1} = p_{\max}\sigma_0 \tag{16.4}$$

and $(\sigma)_{p=1}$ is the tensile stress in the uncracked body at the crack location for $p = 1$. Using the relations (16.3), (16.4) and the expression $K_1 = k\sigma\sqrt{\pi l}$, we can write the formula for l_{cr} in the form

$$l_{cr} = \frac{1}{\pi}\left(\frac{K_{1C}}{k\sigma_{\max}}\right)^2 = \frac{1}{\pi}\left(\frac{K_{1C}}{kp_{\max}\sigma_0}\right)^2. \tag{16.5}$$

As previously described, we consider the initial size of the cracks as a random value $\xi = l_i$. Since these values and also other mechanical parameters are varied between given limits, it is convenient to use truncated probabilistic distributions.

In what follows we will use normal truncated distribution [Ara87] with the following density function $\widehat{f}(x)$:

$$\widehat{f}(x) = 0, \ \xi \leq \xi_1,$$

$$\widehat{f}(x) = 0, \ \xi \geq \xi_1, \tag{16.6}$$

$$\widehat{f}(x) = \frac{\widehat{C}}{\widehat{\sigma}\sqrt{2\pi}} \exp\left[-\frac{(\xi - \widehat{m})^2}{2\widehat{\sigma}^2}\right], \ \xi_1 \leq \xi \leq \xi_2,$$

and an integral distribution function $\widehat{F}(x)$ defined as

$$\widehat{F}(\xi) = 0, \ \xi \leq \xi_1,$$

$$\widehat{F}(\xi) = 1, \ \xi \geq \xi_1, \tag{16.7}$$

$$\widehat{F}(\xi) = \widehat{C}\left[Z(t) - Z(t_1)\right],$$

where ξ_1, ξ_2 are the given limits and $\widehat{m}$, $\widehat{\sigma}^2$ are, respectively, the mathematical expectation and the dispersion of the original (untruncated) distribution. For compactness, we used the following notations:

$$\widehat{C} = \frac{1}{Z(t_2) - Z(t_1)}, \ t = \frac{\xi - \widehat{m}}{\widehat{\sigma}},$$

$$t_1 = \frac{\xi_1 - \widehat{m}}{\widehat{\sigma}}, \ t_2 = \frac{\xi_2 - \widehat{m}}{\widehat{\sigma}}, \tag{16.8}$$

$$Z(t) = \frac{1}{\sqrt{2\pi}} \int_{-\infty}^{t} \exp\left(-\frac{x^2}{2}\right) dx,$$

where $Z(t)$ is the tabulated function of normal distribution having the properties

$$Z(-\infty) = 0, Z(+\infty) = 1, \tag{16.9}$$

$$Z(-t) = 1 - Z(t).$$

16.2 Probabilistic Optimization

Let us introduce a small positive parameter ε and write the limiting stress intensity factor $K_{1\varepsilon} = K_{1C} - \varepsilon, \varepsilon \geq 0$, as discussed before [BRS99, BRS03a, BRS06]. We can write the fracture mechanics criterion as

$$K_1 = K_{1\varepsilon}. \tag{16.10}$$

Consider the problem of finding the design variable h characterizing the geometry of the body that minimizes the mass (or the volume) J of the body:

$$J(h) \rightarrow \min_{h}, \tag{16.11}$$

while satisfying the probabilistic constraint

$$P\{K_1 \leq K_{1\varepsilon}\} \geq 1 - \nu \tag{16.12}$$

and the longevity condition

$$n_{cr} \geq n_0, \tag{16.13}$$

where n_0 is a given minimum value, and ν is a small positive given parameter $(0 < \nu < 1)$. The probability of the value K_1 to be equal or less than the value $K_{1\varepsilon}$ can be evaluated as

$$P\{\xi \leq \xi_{cr}\} = \int_0^{\xi_{cr}} f(\xi)d\xi = F(\xi_{cr}), \tag{16.14}$$

where $f(\xi)$, $F(\xi)$ are the given density and distribution functions for the random variable ξ, and ξ_{cr} is the critical value of the initial crack length $(l_0)_{cr}$ that reaches the value l_{cr} after n_0 loading cycles. The condition (16.12) takes the form

$$F(\xi_{cr}) \geq 1 - \nu. \tag{16.15}$$

Taking into account the monotonicity of the function $F(\xi)$, we can represent the condition (16.15) as

$$\xi_{cr} \geq S(1 - \nu), \tag{16.16}$$

where S is the inverse of the function F, i.e.,

$$S(F(\xi)) = \xi$$

for any ξ.

Let us relate ξ_{cr} to the parameters n_{cr} and l_{cr}. We represent Paris' law (16.1) in the following form:

$$\frac{dl}{l^{\alpha/2}} = \Phi dn, \tag{16.17}$$

$$\Phi \equiv C(k\sigma_0\sqrt{\pi}\Delta p)^{\alpha}, \quad \Delta p = p_{\max} - p_{\min} \tag{16.18}$$

and perform integration over the intervals $[(l_i)_{cr}, l_{cr}]$ for the variable l, and $[0, n_{cr}]$ for the variable n. Taking into account that

$$(l_i) = \xi_{cr} \text{ and } n_{cr} = n_0,$$

we will have

$$\Phi n_0 = \Phi \int_0^{n_0} dn = \int_{\xi_{cr}}^{l_{cr}} \frac{dl}{l^{\alpha/2}} = \frac{2}{\alpha - 2} \left\{ \frac{1}{\xi_{cr}^{(\alpha-2)/2}} - \frac{1}{l_{cr}^{(\alpha-2)/2}} \right\}. \tag{16.19}$$

We arrive at the following relation for ξ_{cr}

$$\frac{1}{\xi_{cr}^{(\alpha-2)/2}} = \frac{\alpha-2}{2}\Phi n_0 + \frac{1}{l_{cr}^{(\alpha-2)/2}} \tag{16.20}$$

On the other hand,

$$\frac{1}{\xi_{cr}} \leq \frac{1}{S(1-v)} \tag{16.21}$$

as follows from the inequality (16.16). Using the relations (16.20) and (16.21), we exclude the value ξ_{cr} and derive the inequality

$$\left(\frac{\alpha-2}{2}\right)\Phi n_0 + \frac{1}{l_{cr}^{(\alpha-2)/2}} \leq \frac{1}{[S(1-v)]^{(\alpha-2)/2}}. \tag{16.22}$$

Let us use the formulas (16.5) and (16.18) for l_{cr}, Φ, and introduce the following notation:

$$a_1 = \left(\frac{\alpha-2}{2}\right)n_0 C\left[\sqrt{\pi}k\left(p_{\max} - p_{\min}\right)\right]^\alpha,$$

$$a_2 = \left(\frac{\sqrt{\pi}kp_{\max}}{K_{1\varepsilon}}\right)^{\alpha-2}, \tag{16.23}$$

$$a_3 = \frac{1}{[S(1-v)]^{(\alpha-2)/2}}.$$

Then we can rewrite the inequality (16.22) as a constraint on the stress σ_0:

$$\Psi \equiv \sigma_0^\alpha + \frac{a_2}{a_1}\sigma_0^{\alpha-2} - \frac{a_3}{a_1} \leq 0. \tag{16.24}$$

The inequality (16.24) can be solved numerically for any $\alpha \in (2,4]$. For particular cases when $\alpha = 3$ or $\alpha = 4$, it can be solved analytically. Consider, for example, the important case when $\alpha = 4$, which is typical for metals. The inequality (16.24) takes the form

$$\Psi \equiv \sigma_0^4 + \frac{a_2}{a_1}\sigma_0^2 - \frac{a_3}{a_1} = \left(\sigma_0^2 - \sigma_{0+}^2\right)\left(\sigma_0^2 - \sigma_{0-}^2\right) \leq 0, \tag{16.25}$$

$$\left(\sigma_0^2\right)_\pm = -\frac{a_2}{2a_1} \pm \sqrt{\frac{a_2^2}{4a_1^2} + \frac{a_3}{a_1}}.$$

Since a_1, a_2, a_3 are all positive, Fig. 16.1 and (16.25) show that the solution of the inequality (16.25) can be written as

$$0 \leq \sigma_0^2 \leq \left(\sigma_0^2\right)_+ \equiv \chi. \tag{16.26}$$

Fig. 16.1 Behavior of the Ψ function

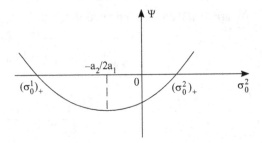

Before considering concrete examples of the solution of the optimization problem, let us make the following note.

In this chapter we supposed implicitly that the fracture is immediately realized when the crack size attains the critical value determined by the quasibrittle fracture criterion, and that the fracture has unstable dynamical character (global destruction). This is typical for these structures. But, of course, there exist other cases. For example, the crack can continue to grow in a stable manner if the derivative of the stress intensity factor with respect to crack length becomes negative (just after attaining the condition $l = l_{cr}$). In some cases, the "critical crack" makes a jump and stable growth is continued. For such cases, the analysis and optimization problems can be generalized, but their solution becomes very complicated.

16.3 Examples of Probabilistic Problems

We can find explicit expression for σ_0 as a function of design variables and loading functions

$$\sigma_0 = \sigma_0(x, q_0(x), h(x))$$

for statically determinate structures; the optimal design can be found using these requirements

$$0 \leq \sigma_0^2(x, q_0(x), h(x)) \leq \chi, \qquad (16.27)$$

$$J = J(h) \rightarrow \min_h. \qquad (16.28)$$

16.3.1 Simply Supported Beam

Consider the optimal design of a beam having a rectangular cross-section with constant width b and variable thickness $h = h(x)$ ($0 \leq x \leq L$). The beam is loaded by given distributed moment $M = pM_0(x)$ (m_0 is a amplitude function) and simply supported at its ends $x = 0$ and $x = L$. For the beam, we will have

$$\sigma_0^2 = \frac{36M_0^2(x)}{b^2h^4(x)}, \; 0 \leq x \leq L, \tag{16.29}$$

$$J = J(h) = b \int_0^L h(x)dx. \tag{16.30}$$

Using these expressions and the requirements (16.29) and (16.30), we find the optimal distribution of the thickness of the beam

$$h(x) = \left(\frac{6|M_0(x)|}{b\sqrt{\chi}}\right)^{1/2}. \tag{16.31}$$

Suppose now that we have an additional size constraint imposed on the design variable h:

$$h(x) \geq h_0,$$

where h_0 is a given constant. The problem is formulated as

$$b \int_0^L h(x)dx \to \min_h, \tag{16.32}$$

$$h(x) \geq h_0, \; h(x) \geq \left(\frac{6|M_0(x)|}{b\sqrt{\chi}}\right)^{1/2}$$

and the solution is

$$h(x) = \max \left\{ h_0, \left(\frac{6|M_0(x)|}{b\sqrt{\chi}}\right)^{1/2} \right\}. \tag{16.33}$$

The relative gain of the optimal design (16.31) referred to the total mass of the beam with constant thickness

$$h_C = \left(\frac{6}{b\sqrt{\chi}} \max_{x\in[0,L]} |M_0(x)|\right)^{1/2} \tag{16.34}$$

is

$$\beta = \frac{J_C - J_{opt}}{J_C} = 1 - \frac{1}{L} \int_0^L \sqrt{\widetilde{M}_0(x)}dx, \tag{16.35}$$

where

$$\widetilde{M}_0(x) = \frac{|M_0(x)|}{M_*}, \; M_* = \max_{x\in[0,L]} |M_0(x)|. \tag{16.36}$$

For example, if the beam is loaded by a point force concentrated at the center point $x = L/2$, we will have

$$M_0(x) = \frac{x}{2}, \ 0 \le x \le \frac{L}{2}, \tag{16.37}$$

$$M_* = \frac{L}{4}$$

and the gain is

$$\beta = 1 - \frac{2}{L} \int\limits_0^{L/2} \sqrt{\widetilde{M}_0(x)}dx \sim 33\%. \tag{16.38}$$

Here we take into account the symmetry of the problem with respect to the point $x = L/2$ and use dimensionless variables. The optimal design (16.33) of the problem (16.32) is characterized by the gain of optimization that satisfies the inequality $0 \le \beta \le 33\%$ and decreases from 33% down to 0 when the problem parameter h_0 increases from 0 up to the value $h_0 = h_c$.

16.3.2 Cantilever Beam

As another example, consider optimal design for a cantilever beam loaded by distributed bending moment $M = pM_0(x)$, $0 \le x \le L$, and a tensile force $Q = pQ_0$ applied to the free end $x = L$ and directed in the longitudinal direction. The end $x = 0$ is clamped. Assuming that the beam has a rectangular cross-section of constant width and a variable thickness $h = h(x)$, we take the function $h = h(x)$ as an unknown design variable and use the following expression for the maximum stress

$$\sigma_0 = \frac{Q_0}{bh} + \frac{6|M_0(x)|}{bh^2}. \tag{16.39}$$

The minimized volume of the beam is given by the corresponding expression in (16.30). From the requirement (16.27), we will have

$$h^2 - \frac{Q_0}{b\sqrt{\chi}}h - \frac{6|M_0(x)|}{b\sqrt{\chi}} \ge 0. \tag{16.40}$$

Optimal design, which realizes the minimum of the functional (16.30) (volume of the beam), under constraint (16.26), (16.27) can be written as

$$h(x) = \frac{Q_0}{2b\sqrt{\chi}} + \left(\frac{Q_0^2}{4b^2\chi} + \frac{6|M_0(x)|}{b\sqrt{\chi}} \right)^{1/2}. \tag{16.41}$$

For statically determinate frames loaded by transverse forces, the optimal design is also given by the formula (16.41).

Fig. 16.2 Thickness
distribution for optimal frame

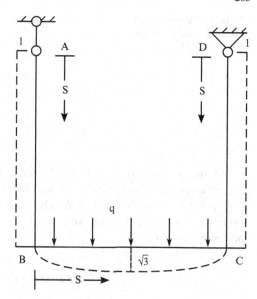

16.3.3 Optimal Design of a Frame

Consider the optimal design for the frame shown in Fig. 16.2. Elements of the frame have equal lengths $L = 2$ (dimensionless variables are used) and constant cross-section width and variable thickness. The thickness distribution is taken as a design variable. The element BC is loaded by uniform distributed transverse force $q = pq_0(s)$, $q_0(s) = 1$. The coordinate s ($0 \leq s \leq 2$) runs from A to B, from B to C, and from D to C. Bending moments $M = pM_0(s)$ and normal forces $R = pR_0(s)$ induced by the transverse load q are given by the following expressions:

$$AB: \quad M_0(s) = 0, \quad R_0(s) = 1, \ 0 \leq s \leq 2,$$
$$BC: \quad M_0(s) = s\left(1 - \frac{s}{2}\right), \quad R_0(s) = 0, \ 0 \leq s \leq 2, \qquad (16.42)$$
$$DC: \quad M_0(s) = 0, \quad R_0(s) = 1, \quad 0 \leq s \leq 2.$$

The optimal thickness distribution, given by the formula (16.41) (where $Q_0 = R_0$, $b\sqrt{\chi} = 1$), is shown by dashed lines.

16.3.4 Clamped-Simply Supported Beam

As an example of optimal design of a statically indeterminate structure, we consider a three-layered beam simply supported at the left end ($x = 0$), and clamped at the right end ($x = l$). The beam has a rectangular cross-section with constant

Fig. 16.3 Cyclic loading of
the statically indeterminate
three-layered beam

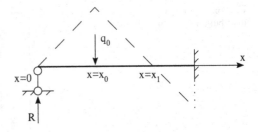

width b and thickness H of the internal layer, and variable thickness $h = h(x)$ of the
external layers. It is loaded by the concentrated force $q = pq_0$ ($p_{min} \leq p \leq p_{max}$)
at the point $x = x_0$, $0 < x_0 < L$. The thickness distribution $h(x)$ is taken as a
design variable. The following expressions can be taken for the moment of inertia I
and for the stress σ_0:

$$I = c_1 h(x), \quad c_1 = \frac{bH^2}{2},$$

$$\sigma_0 = c_2 \frac{M_0}{h}, \quad c_2 = \frac{1}{bH}. \tag{16.43}$$

Let us denote the unknown reaction force acting at the left end (see Fig. 16.3) by R
and write the distribution of the moment $M_0(x)$ in the following manner:

$$M_0 = Rx, \quad x \in [0, x_0],$$

$$M_0 = q_0 x_0 - x(q_0 - R), \quad x \in [x_0, L]. \tag{16.44}$$

The problem is reduced to the minimization of the volume of the protective material
under the inequalities imposed on σ_0, i.e.,

$$J = 2b \int_0^L h(x) dx \to \min_h,$$

$$h(x) \geq \frac{c_2 M_0(x)}{\sqrt{\chi}} = \frac{c_2 Rx}{\sqrt{\chi}}, \quad 0 \leq x \leq x_0,$$

$$h(x) \geq \frac{c_2 M_0(x)}{\sqrt{\chi}} = \frac{c_2 [q_0 x_0 - x(q_0 - R)]}{\sqrt{\chi}}, \quad x_0 \leq x \leq x_1, \tag{16.45}$$

$$h(x) \geq \frac{c_2 M_0(x)}{\sqrt{\chi}} = -\frac{c_2 [q_0 x_0 - x(q_0 - R)]}{\sqrt{\chi}}, \quad x_1 \leq x \leq L.$$

Here $x_1 = q_0 x_0 / (q_0 - R)$ is the point at which $M_0(x_1) = 0$. It is possible to show
that the solution of problem (16.45) can be written as

$$h(x) = \frac{c_2 R x}{\sqrt{\chi}}, \ 0 \leq x \leq x_0,$$

$$h(x) = \frac{c_2 |q_0 x_0 - x (q_0 - R)|}{\sqrt{\chi}}, \ x_0 \leq x \leq L. \tag{16.46}$$

To find the unknown value R in (16.46), we use the principle of virtual work [Tim56]

$$\int_0^L \frac{M_0 M_1}{EI} dx = 0, \tag{16.47}$$

where E is Young's modulus, and M_1 is the moment created by the unit reaction force $R = 1$ (i.e. $M_1(x) = x$). The relations (16.43) and (16.47) yield

$$\int_0^{x_0} \frac{M_0 M_1}{h} dx + \int_{x_0}^{x_1} \frac{M_0 M_1}{h} dx + \int_{x_1}^{L} \frac{M_0 M_1}{h} dx = 0. \tag{16.48}$$

Substitute of the expressions (16.44) and (16.46) into Eq. (16.48) gives the following value for reaction force:

$$R = q_0 \left(1 - \frac{\sqrt{2} x_0}{L} \right). \tag{16.49}$$

16.4 Moment Constraints

As previously, we will consider one crack parameter (initial size of the crack l_i) as a random variable ξ with known density function $f(\xi)$, whereas the other crack parameter will be considered as given or will be found with the help of the guaranteed approach of game theory (design for the worst case). The probability density distribution $f(\xi)$ for the variable ξ permits us, at least in principle, to determine the moments of the random quantity n_{cr} and, in particular, its mathematical expectation and dispersion (variance). This permits us, in turn, to control the probability of violating the assigned longevity constraint $n_{cr} \geq n_0$ and to study the following problem of optimal design, consisting in finding a design variable h that satisfies the system of inequalities

$$\widehat{n}_{cr} = \mathrm{E}\,(n_{cr}) = \int_0^{\xi_0} n_{cr}\,(\xi, h)\,f(\xi) \geq n_0, \tag{16.50}$$

$$D\,(n_{cr}) = \mathrm{E}\left((n_{cr} - \widehat{n}_{cr})^2 \right) = \int_0^{\xi_0} (n_{cr}\,(\xi, h) - \widehat{n}_{cr})^2\,f(\xi)d\xi \leq D_0 \tag{16.51}$$

and minimizes the functional $J(h)$ (volume or weight of the structure or structural element), while taking care of the equations of the theory of elasticity or the theory of strength of materials. Here $E(n_{cr})$ and $D(n_{cr})$ denote the mathematical expectation and dispersion (variance) for a random variable n_{cr} and D_0 denotes a relatively small parameter (with respect to the given number n_0) that is supposed to be given. The given value $\xi_*(\xi_* < L)$ determines the interval $[0, \xi_*]$ of variation of the random variable ξ. The inequalities (16.50) and (16.51) imply that some conditions must accompany the choice of the design variable h. The longevity constraint given in (16.13) must be satisfied in a "probabilistic" sense, and the dispersion of the quantity n_{cr} must not exceed some chosen relatively small value (small with respect to n_0).

We find n_{cr} by integrating (16.1). Performing elementary transformations, we will have

$$n_{cr} = \frac{b}{\sigma_0^\alpha l_i^{\alpha/2-1}} \left[1 - \left(b_2 l_i \sigma_0^2 \right)^{\alpha/2-1} \right],$$

$$b_1 = \left[\left(\frac{\alpha}{2} - 1 \right) C k^\alpha \pi^{\alpha/2} \left(\Delta p \right)^\alpha \right]^{-1}, \quad b_2 = \frac{k^2 p_{\max}}{K_{1\varepsilon}^2}. \tag{16.52}$$

As can be seen from the expressions (16.52), the value n_{cr} depends on σ_0 and, consequently, the dependence of the critical number of cycles n_{cr} on the geometric design parameters that define the shape of the body, is realized by means of the dependence of the considered normal stresses on these parameters.

The expressions for $E(n_{cr})$, $\left(n_{cr}^2 \right)$, and $D(n_{cr})$ have the form

$$\widehat{n}_{cr} = E(n_{cr}) = \frac{b_1 E \left(\xi^{(2-\alpha)/2} \right)}{\sigma_0^\alpha} - \frac{b_1 b_2^{(\alpha-2)/2}}{\sigma_0^2},$$

$$E(n_{cr}^2) = \frac{b_1^2 E \left(\xi^{2-\alpha} \right)}{\sigma_0^{2\alpha}} - \frac{2 b_1 b_2^{(\alpha-2)/2} E \left(\xi^{(\alpha-2)/2} \right)}{\sigma_0^{\alpha+2}} - \frac{b_1^2 b_2^{\alpha-2}}{\sigma_0^4}, \tag{16.53}$$

$$D(n_{cr}) = E(n_{cr}^2) - (E(n_{cr}))^2 = \frac{b_1^2}{\sigma_0^{2\alpha}} \left[E(\xi^{2-\alpha}) - E^2 \left(\xi^{(2-\alpha)/2} \right) \right] \geq 0.$$

Using these expressions and the problem constraints (16.50) and (16.51), we derive the following system of inequalities:

$$\sigma_0^\alpha + \lambda_1 \sigma_0^{\alpha-2} - \lambda_2 \leq 0, \quad \sigma_0^{2\alpha} \geq \lambda_3,$$

$$\lambda_1 = \frac{b_1 b_2^{(\alpha-2)/2}}{n_0}, \quad \lambda_2 = \frac{b_1 E \left(\xi^{(2-\alpha)/2} \right)}{n_0}, \tag{16.54}$$

$$\lambda_3 = \frac{b_1^2}{D_0} \left[E(\xi^{2-\alpha}) - E^2 \left(\xi^{(2-\alpha)/2} \right) \right]$$

determine the admissible values of σ_0. If $\alpha = 4$, we find

$$\lambda_3^{1/4} \leq \sigma_0^2 \leq (\sigma_0^2)_+ = \mu,$$

$$(\sigma_0^2)_+ = -\frac{1}{2}\lambda_1 + \sqrt{\frac{\lambda_1^2}{4} + \lambda_2}, \tag{16.55}$$

where $(\sigma_0^2)_+$ is the positive root of the following equation:

$$\sigma_0^\alpha + \lambda_1\sigma_0^{\alpha-2} - \lambda_2 = 0.$$

Note that the solution of the optimization problem exists if the problem parameters satisfy the inequality $\lambda_3^{1/3} \leq \mu$. Determination of the optimal design for statically determinate and statically indeterminate beams and frames can be performed analytically as in Sections 16.2 and 16.3. Taking into account the character of the dependence of σ_0 on the design variable h for the problem presented in Section 16.3, we can determine that the right-hand side inequality in (16.55) is realized as a rigorous equality for the optimal solution of corresponding problems with the constraints (16.50) and (16.51). Thus all solutions presented in Section 16.3 can be considered as solutions of the problem with moment constraints if we substitute the parameter μ into the corresponding expression instead of the parameter χ.

In some cases, we have additional probabilistic information concerning problem parameters and can formulate probabilistic optimization problem with several random variables. Consider, for example, the optimal design of a cantilever three-layered beam taking not only the size $l_i = \xi$ ($0 \leq \xi \leq \xi_0$) of the initial surface crack, but also the location $x = \eta$ ($0 \leq \eta \leq L$) of the crack as a random variable. We assume that the random variables are independent and characterized by the given density probability functions $f(\xi)$ and $g(\eta)$. Suppose that the beam is clamped at the left end ($x = 0$) and has a rectangular cross-section with constant thickness H of the internal layer and variable thickness $h = h(x)$ ($0 \leq x \leq L$) of the external layers ($h \ll H$). The beam is loaded by cyclic forces with distributed moment $pM_0(x)$ ($0 \leq p_{min} \leq p \leq p_{max}$). The mathematical expectation of the critical number of cycles can be written as

$$\widehat{n}_{cr} = \mathrm{E}(n_{cr}) = \int_0^L \int_0^{\xi_0} n_{cr}(\xi, \eta)\, f(\xi)g(\eta)d\xi d\eta = r_1 J_1 - r_2 J_2, \tag{16.56}$$

where the parameters r_1, r_2 and the integrals J_1, J_2 are given by the formulas

$$r_1 = b_1\mathrm{E}(\xi^{(2-\alpha)/2}), \quad r_2 = b_1 b_2^{(\alpha-2)/2},$$

$$J_1 = \int_0^L \frac{g(\eta)d\eta}{\sigma_0^\alpha}, \quad J_2 = \int_0^L \frac{g(\eta)d\eta}{\sigma_0^2}. \tag{16.57}$$

The problem consists in finding the thickness distribution $h = h(x)$ to minimize the volume of external layers under a constraint on the mean value of the critical numbers of cycles:

$$J_{opt} = \min_h J(h) = \min_h c_3 \int_0^L h(x)dx \qquad (16.58)$$

$$r_1 J_1 - r_2 J_2 \geq n_0 \qquad (16.59)$$

where $c_3 = 2b$. For simplification, we consider only the moment constraint and the case $\alpha = 3$. It is possible to show that the minimum of the functional (16.58) under the constraint (16.59) is reached when the rigorous equality is realized in (16.59). In this case, the augmented Lagrange functional has the from

$$J^a = \int_0^L \left[c_3 h - \mu \left(\frac{d_1 g h^3}{|M_0|^3} - \frac{d_2 g h^2}{M_0^2} \right) \right] dx \qquad (16.60)$$

$$d_1 = \frac{r_1}{c_2^3}, \ d_2 = \frac{r_2}{c_2^2}.$$

The necessary extremum condition gives the following expression for the optimal thickness distribution:

$$h = \frac{d_2 |M_0|}{3 d_1} + \left(\frac{d_2^2 M_0^2}{9 d_1^2} + \frac{|M_0|^3 c_3}{3 \mu d_1 g} \right)^{1/2}, \qquad (16.61)$$

where μ is the Lagrange multiplier determined with the help of the equality

$$r_1 J_1 - r_2 J_2 = n_0.$$

Let us consider, for example, the following data (forces are in N and lengths in m), where $f(\xi)$ and $g(\eta)$ are truncated normal distributions, and $\widehat{m}_\xi, \widehat{m}_\eta, \widehat{\sigma}_\xi, \widehat{\sigma}_\eta$ are mathematical expectations and standard deviations of the corresponding original

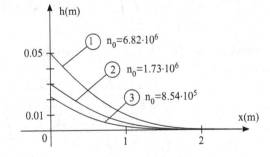

Fig. 16.4 Optimal thickness distribution of external layers when crack size and location are random

(non truncated) normal density functions: $\xi_1 = 0$, $\xi_2 = 10^{-8}$, $\widehat{m}_\xi = 10^{-8}/2$, $\widehat{\sigma}_\xi = 10^{-9}$, $\xi_0 = 10^{-6}$, $\eta_1 = 0$, $\eta_2 = 2$, $\widehat{m}_\eta = 1$, $\widehat{\sigma}_\eta = 1$, $\alpha = 3$, $L = 2$, $b = 0.1$, $H = 0.3$, $k = 1.12$, $K_{1\varepsilon} = 27 \cdot 10^6$, $C = 3.1 \cdot 10^6$ $\Delta p = 10^2$, $M_0(x) = ((L - x)/2) \cdot 10^{-5}$. Varying the parameter n_0, we can find the corresponding function $h(x)$. The resultant solution is presented in Fig. 16.4 for the cases $n_0 = 6.82 \cdot 10^6$ (case 1); $n_0 = 1.73 \cdot 10^6$ (case 2) and $n_0 = 8.54 \cdot 10^5$ (case 3).

Chapter 17
Mixed Probabilistic-Guaranteed Optimal Design

This chapter deals with the mixed probabilistic-guaranteed approach to optimal design of quasi-brittle membrane shells. Special attention is devoted to different problem formulations and analytical methods for their solution. Optimal thickness distributions are presented for various axisymmetric membrane shells. The presentation follows research results of [BRS03b].

17.1 Probabilistic-Guaranteed Optimization

Consider a shell that is a surface of revolution, and loaded by axisymmetric forces in the meridian planes. Intensities of the external loads which act in the direction normal and tangential to the meridian, are denoted by q_n and q_φ. The position of a meridian is defined by the angle θ, measured from some datum meridian plane, and the position of a parallel circle is defined by the angle φ, made by the normal to the surface and the axis of rotation (Fig. 17.1).

The radius of the parallel circle and the radii of curvature are denoted by r, r_φ and r_θ, respectively. The equilibrium equations and the expressions for the normal stresses σ_φ, σ_θ are written in the following form [Tim56, TW59]:

$$\frac{N_\varphi}{r_\varphi} + \frac{N_\theta}{r_\theta} = q_n, \; 2\pi r N_\varphi \sin \varphi + R = 0,$$

$$\sigma_\varphi = N_\varphi(\varphi)/h(\varphi), \; \sigma_\theta = N_\theta(\varphi)/h(\varphi), \tag{17.1}$$

where $h = h(\varphi)$ is the shell thickness, which can be varied in the meridian direction, N_φ, N_θ are the normal membrane forces.

It is assumed that a through the thickness crack can arise in the shell. The crack is rectilinear, and its length is larger than h but small with respect to the characteristic size of the shell. The crack size l and its location and orientation, given by the coordinate φ_C, θ_C of the crack midpoint, and the angle α setting the crack inclination with respect to the meridian, are not fixed beforehand. For the stress intensity factor K_1 we will use the expression $K_1 = \sigma_n^0 \sqrt{\pi l/2}$ if $\sigma_n^0 \geq 0$ and $K_1 = 0$ if $\sigma_n^0 < 0$, where σ_n^0 is the normal stress in the uncracked shell at the place of the crack loca-

N.V. Banichuk and P.J. Neittaanmäki, *Structural Optimization with Uncertainties*, Solid Mechanics and Its Applications 162, DOI 10.1007/978-90-481-2518-0_1,

Fig. 17.1 Portion of the
axisymmetric shell above
parallel circle AB

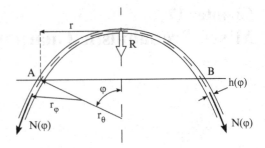

tion. Note that the subscript n means that the stress acts in the direction normal to
the crack banks; this is correct when the crack is distant from the shell boundaries.

In this discussion, we consider one crack parameter as a random variable ξ with
known density function $f(\xi)$, while the other crack parameters are considered as
components of vector ω belonging to the admissible set Λ_ω, i.e.

$$\omega \in \Lambda_\omega \tag{17.2}$$

found from the guaranteed approach of game theory (design for the worst case). In
general terms, the stress intensity factor depends on the random variable ξ, vector ω
and the design variable h, i.e.

$$K_1 = K_1(\xi, \omega, h) \tag{17.3}$$

For definiteness, we consider the length of the crack $\xi (l = \xi)$ and the values α and
φ_C as components of the vector ω.

The probability density distribution $f(\xi)$ for the random variable ξ permits us,
at least in principle, to determine the moments of the random quantity K_1 and,
in particular, its mathematical expectation and dispersion (variance). This in turn
permits us to control the probability of violating the assigned strength constraint
$K_1 \leq K_{1\varepsilon}$ and to study the following problem of optimal design. Find a function h
that satisfies the inequality imposed on the mathematical expectation

$$\max_{\omega} \widehat{K}_1 = \max_{\omega} \mathrm{E}(K_1) = \max_{\omega} \int_0^\infty K_1(\xi, \omega, h) f(\xi) d\xi \leq K_{1\varepsilon} \tag{17.4}$$

the inequalities imposed on the dispersion

$$\max_{\omega} D(K_1) = \max_{\omega} \mathrm{E}\left(\left(K_1 - \widehat{K}_1\right)^2\right) = \max_{\omega} \int_0^\infty \left(K_1(\xi, \omega, h) - \widehat{K}_1\right)^2 f(\xi) d\xi \leq \delta \tag{17.5}$$

and thickness distribution

$$h(\varphi) \geq h_0 \tag{17.6}$$

and minimizes the functional $J(h)$ (volume of the material of the shell or weight of the shell)

$$J = J(h) \to \min_{h}, \tag{17.7}$$

$$J(h) = 2\pi \int_{\varphi_0}^{\varphi_f} h r_\varphi r_\theta \sin \varphi \, d\varphi. \tag{17.8}$$

Here h_0, δ are given positive constants and φ_0, φ_f are given limits for the independent variable φ ($\varphi_0 \leq \varphi \leq \varphi_f$). Thus, the brittle fracture mechanics constraint $K_1 \leq K_{1\varepsilon}$ must be satisfied in "the average" sense, and the deviation of quantity $K_1 - \widehat{K}_1$ must not exceed some sufficiently small positive number.

17.2 Analytic Solution

Using the explicit representation for K_1 and noting that the extremal values of σ_n^0 with respect to the angle of inclination α are realized for $\alpha = 0$ (meridian direction) and for $\alpha = \pi/2$ (parallel direction), we find that the maximum of K_1 with respect to α is attained when α takes one of two values $\alpha = 0$ (axial crack) or $\alpha = \pi/2$ (peripheral crack). Thus we have

$$\max_{\omega} K_1 = \max \left\{ \max_{\varphi} (K_1)_{\alpha=0}, \ \max_{\varphi} (K_1)_{\alpha=\pi/2} \right\}, \tag{17.9}$$

where

$$(K_1)_{\alpha=0} = \sqrt{\frac{\pi \xi}{2}} (\sigma_n)_{\alpha=0} = \frac{\sqrt{\xi}}{h} \Phi_1, \tag{17.10}$$

$$\Phi_1 = r_\theta \sqrt{\frac{\pi}{2} \left(\frac{R}{2\pi r r_\varphi \sin \varphi_c} + q_n \right)}, \tag{17.11}$$

$$(K_1)_{\alpha=\pi/2} = \sqrt{\frac{\pi \xi}{2}} (\sigma_n)_{\alpha=\pi/2} = \frac{\sqrt{\xi}}{h} \Phi_2, \tag{17.12}$$

$$\Phi_2 = \frac{R}{2\sqrt{2\pi} r \sin \varphi_c}. \tag{17.13}$$

Note here that

$$(\sigma_n)_{\alpha=0} = \sigma_\theta, \ (\sigma_n)_{\alpha=\pi/2} = \sigma_\varphi.$$

We can rewrite the expression (17.9) in the form

$$\max_{\omega} K_1 = \frac{\sqrt{\xi}}{h} \max \left\{ \max_{\varphi_c} \Phi_1 (\varphi_c), \ \max_{\varphi_c} \Phi_2 (\varphi_c) \right\}. \tag{17.14}$$

For the mean value of K_1 (mathematical expectation $E(K_1)$) and the variance $D(K_1)$, we will have

$$E\left((K_1)_{\alpha=0}\right) = \int_0^\infty ((K_1)_{\alpha=0}) \, f(\xi)d\xi = \frac{\Phi_1}{h} E\left(\sqrt{\xi}\right), \tag{17.15}$$

$$E\left((K_1)_{\alpha=\pi/2}\right) = \int_0^\infty ((K_1)_{\alpha=\pi/2}) \, f(\xi)d\xi = \frac{\Phi_2}{h} E\left(\sqrt{\xi}\right), \tag{17.16}$$

$$D\left((K_1)_{\alpha=0}\right) = \int_0^\infty ((K_1)_{\alpha=0} \, E\,(K_1)_{\alpha=0})^2 \, f(\xi)d\xi = \frac{\Phi_1^2}{h^2} D\left(\sqrt{\xi}\right), \tag{17.17}$$

$$D\left((K_1)_{\alpha=\pi/2}\right) = \int_0^\infty ((K_1)_{\alpha=\pi/2} \, E\left((K_1)_{\alpha=\pi/2}\right))^2 \, f(\xi)d\xi = \frac{\Phi_2^2}{h^2} D\left(\sqrt{\xi}\right). \tag{17.18}$$

Thus, maximum values of $E(K_1)$ and $D(K_1)$ are given by the expressions

$$\max_\omega E(K_1) = E\left(\sqrt{\xi}\right) \left\{ \max \left\{ \max_{\varphi_c} \frac{\Phi_1}{h}, \ \max_{\varphi_c} \frac{\Phi_2}{h} \right\} \right\}, \tag{17.19}$$

$$\max_\omega D(K_1) = D\left(\sqrt{\xi}\right) \left\{ \max \left\{ \max_{\varphi_c} \frac{\Phi_1^2}{h^2}, \ \max_{\varphi_c} \frac{\Phi_2^2}{h^2} \right\} \right\}. \tag{17.20}$$

Using the expressions (17.15) through (17.20), we can present the fracture mechanics constraints (17.4) and (17.5) as a system of four inequalities

$$E\left(\sqrt{\xi}\right) \max_{\varphi_c} \left(\frac{\Phi_i}{h}\right) \le K_{1\varepsilon}, \ i = 1, 2, \tag{17.21}$$

$$D\left(\sqrt{\xi}\right) \max_{\varphi_c} \left(\frac{\Phi_i^2}{h^2}\right) \le \delta, \ i = 1, 2.$$

The problem consists in finding an admissible thickness distribution that minimizes the functional J, while satisfying the geometrical condition (17.6) and the strength constraints (17.21). The solution of this problem is written in explicit form:

$$h(\varphi) = \max \{h_0, \Psi_{E1}(\varphi), \Psi_{E2}(\varphi), \Psi_{D1}(\varphi), \Psi_{D2}(\varphi)\}, \tag{17.22}$$

where

$$\Psi_{Ei} = \frac{\Phi_i}{K_{1E}} E\left(\sqrt{\xi}\right), \ i = 1, 2, \tag{17.23}$$

$$\Psi_{Di} = \frac{\Phi_i}{\sqrt{\delta}} \sqrt{D(\xi)}, \ i = 1, 2$$

represent given positive functions of the angle φ. Let us introduce the problem parameter

$$\lambda = \frac{\sqrt{\delta} E\left(\sqrt{\xi}\right)}{K_{1\varepsilon} \sqrt{D\left(\sqrt{\xi}\right)}}. \tag{17.24}$$

The parameter λ represent the ratio between h calculated from the first expression (17.23) and the second expression (17.23), that is

$$\lambda = \Psi_{Ei} / \Psi_{Di}, \quad i = 1, 2.$$

If $\lambda \geq 1$, then the value of h calculated using the first expression (17.23) is the largest so it represents the optimal solution, which is given by the formula

$$h(\varphi) = \max\{h_0, \Psi_{E1}(\varphi), \Psi_{E2}(\varphi)\}. \tag{17.25}$$

If $\lambda < 1$, then the value of h calculated using the second expression (17.23) is the largest, so the optimal solution takes the form

$$h(\varphi) = \max\{h_0, \Psi_{D1}(\varphi), \Psi_{D2}(\varphi)\}. \tag{17.26}$$

17.3 A Toroidal Shell

For unclosed shells supported usually by some rings or other arrangements against circumferential extension, some bending will occur near the supports; the edge effect is very localized and the edge zone with meaningful moments is relatively small. At a certain distance from the boundary, we can use membrane theory with satisfactory accuracy [TW59]. Noting this, we construct the optimal design for such cases up to a small zone near the boundary.

Consider a toroidal shell obtained by rotating a circle of a radius a about a vertical axis (Fig. 17.2) distant b from the center of the circle. The shell is subject to a uniform pressure p. The forces N_φ, N_θ are written as

$$N_\varphi = \frac{pa(r+b)}{2r}, \ N_\theta = \frac{pa}{2}.$$

Fig. 17.2 Optimal thickness
distribution of toroidal tank

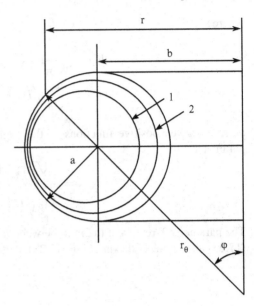

Noting that $N_\varphi > N_\theta$, and, consequently,

$$\Phi_2 = \sqrt{\frac{\pi}{2}N_\varphi} > \Phi_1 = \sqrt{\frac{\pi}{2}N_\theta} \tag{17.27}$$

we have the optimal thickness distribution in the following form:

$$h(\varphi) = \max\{h_0, \Psi_{E2}(\varphi), \Psi_{D2}(\varphi)\}. \tag{17.28}$$

Here, we used the inequalities

$$\Psi_{E2} > \Psi_{E1} \text{ and } \Psi_{D2} > \Psi_{D1}.$$

If $\lambda \geq 1$, then the optimal solution can be written as

$$h(\varphi) = \max\{h_0, \Psi_{E2}(\varphi)\} = \max\left\{h_0, \frac{E\left(\sqrt{\xi}\right)}{K_{1\varepsilon}}\sqrt{\pi/2}\frac{pa(r+b)}{2r}\right\}. \tag{17.29}$$

If $\lambda < 1$, then the optimal solution is

$$h(\varphi) = \max\{h_0, \Psi_{D2}(\varphi)\} = \max\left\{h_0, \frac{\sqrt{D\left(\sqrt{\xi}\right)}}{\sqrt{\delta}}\sqrt{\pi/2}\frac{pa(r+b)}{2r}\right\}. \tag{17.30}$$

For both cases presented in (17.29) and (17.30), the optimal thickness (shown respectively by dashed lines 1 and 2 in Fig. 17.2) decreases when a radius r increases.

17.4 A Shell Loaded by Forces at the Free Ends

Consider a shell of revolution shown in Fig. 17.1. The shell is loaded by axisymmetric forces applied to its free ends; the resultant of forces (applied to the parallel circle) acting in the x-direction is denoted by P_0.

The distance $r = r(x)$ from the axis of revolution to a point of the middle surface is taken as the basic geometrical variable describing the shape of the middle surface, and is supposed to be given. The expressions for the quality functional (volume of the shell material) and the stresses are

$$J = 2\pi \int_0^L rh \left(1 + \left(\frac{dr}{dx}\right)\right)^{1/2} dx, \tag{17.31}$$

$$\sigma_\varphi = \frac{P_0}{2\pi h r} \sqrt{1 + \left(\frac{dr}{dx}\right)^2}, \quad \sigma_\theta = \frac{P_0}{2\pi} \frac{d^2 r/dx^2}{\sqrt{1 + (dr/dx)^2}}.$$

The fracture mechanics constraints are written in the form

$$E\left(\sqrt{\xi}\right) \max_{0 \le x \le L} \left(\frac{\Phi_i}{h}\right) \le K_{1\varepsilon}, \quad D\left(\sqrt{\xi}\right) \max_{0 \le x \le L} \left(\frac{\Phi_i^2}{h^2}\right) \le \delta,$$

$$\Phi_1 = \frac{P_0 \left(d^2 r/dx^2\right)}{2\sqrt{2\pi} \sqrt{1 + (dr/dx)^2}}, \quad \Phi_2 = \frac{P_0}{2\sqrt{2\pi} r} \sqrt{1 + \left(\frac{dr}{dx}\right)^2}.$$

If $\lambda \ge 1$, the optimum thickness distribution is given by

$$h(x) = \max \left\{ h_0, \frac{E\left(\sqrt{\xi}\right)}{K_{1\varepsilon}} \Phi_1(x), \frac{E\left(\sqrt{\xi}\right)}{K_{1\varepsilon}} \Phi_2(x) \right\}, \quad 0 \le x \le L; \tag{17.32}$$

if $\lambda < 1$, then

$$h(x) = \max \left\{ h_0, \frac{\sqrt{D\left(\sqrt{\xi}\right)}}{\sqrt{\delta}} \Phi_1(x), \frac{\sqrt{D\left(\sqrt{\xi}\right)}}{\sqrt{\delta}} \Phi_2(x) \right\}, \quad 0 \le x \le L. \tag{17.33}$$

To illustrate the optimal solution (17.32) and (17.32), we suppose $\lambda \geq 1$ and the shape of the middle surface is conical; the function $r(x)$ is linear: $r(x) = a + kx$. In this case, the optimal thickness distribution is given by the formula

$$h(x) = \max \left\{ h_0, \frac{\mathrm{E}\left(\sqrt{\xi}\right)}{K_{1\varepsilon}} \frac{P_0 \sqrt{\pi(1 + k^2)/2}}{2\pi(a + kx)} \right\}, \quad 0 \leq x \leq L.$$

References

[AB73] G. Augusti and A. Baratta. Theory of probability and limit analysis of structures under multi-parameter loading. In Foundations of Plasticity (Papers, International Symposium, Warsaw, 1972), A. Sawczuk, ed., Noordhoff, Leyden, 347–364, 1973

[ABC84] G. Augusti, A. Baratta, and F. Casciati. Probabilistic Methods in Structural Engineering. Chapman and Hall, London, 1984

[AC79] G. Augusti and F. Casciati. Further studies on structural design for maximum expected utility. In Proceedings of the 3rd International Conference on Applications of Statistics and Probability in Soil and Structural Engineering, Sidney, 2:686–700, 1979

[AGRZ84] E. Atrek, R. H. Gallagher, K. M. Ragsdell, and O. Z. Zienkiewicz, eds., New Directions in Optimum Structural Design, John Wiley & Sons Ltd, Chichester, 1984

[Ara87] A. M. Araslanov. Analysis of Structural Elements of Given Reliability under Random Action. Mashinostroenie, Moscow, 1987, in Russian

[Arm83] J.-L. Armand. Non homogeneity and anisotropy in structural design. In Optimization Methods in Structural Design, H. Eschenauer and N. Olhoff, eds., Bibliographisches Institut, Mannheim, 256–263, 1983

[Aro89] J. S. Arora. Introduction to Optimum Design, McGraw – Hill Book Company, New York, 1989

[Ban07] N. V. Banichuk. Optimization of axisymmetric membrane shells. Journal of Applied Mathematics and Mechanics, 71(4): 527–535, 2007

[Ban75] N. V. Banichuk. Game problems in the theory of optimal design. In Proceedings of IUTAM Symposium on Optimization in Structural Design, A. Sawczuk and Z. Mroz, eds., Springer-Verlag, Berlin, 111–121, 1975

[Ban76] N. V. Banichuk. Minimax approach to structural optimization problems. Journal of Optimization Theory and Applications, 20(1): 111–127, 1976

[Ban81] N. V. Banichuk. Optimization problems for elastic anisotropic bodies. Archives of Mechanics. 33: 347–363, 1981

[Ban83] N. V. Banichuk. Problems and Methods of Optimal Structural Design. Number 26 in Mathematical Concepts and Methods in Science and Engineering. Plenum Press, New York, 1983

[Ban84] N. V. Banichuk. Application of perturbation method to optimal design of structures. In New Directions in Optimum Structural Design, E. Atrek, R. H. Gallagher, K. M. Ragsdell, and O. C. Zienkiewicz, eds., John Wiley & Sons Ltd, Chichester, 231–248, 1984

[Ban90] N. V. Banichuk. Introduction to Optimization of Structures. Springer-Verlag, New York, 1990

[Ban97] N. V. Banichuk. Free boundaries optimization under fracture mechanics constraints. Universității "Ovidius" Constanța. Analele Științifice. Seria Matematică, 5(1): 13–19, 1997

[Ban98] N. V. Banichuk. Optimal design of quasi-brittle elastic bodies with cracks. Mechanics Based Design of Structures and Machines, 26(4): 365–376, 1998

[Ban99a] N. V. Banichuk. Asymptotic approach to optimal structural design with brittle-fracture constraints (Part 1: Prototype problem). In Mechanics of Composite Materials and Structures, C. A. Mota Soares et al., eds., Kluwer Academic Publishers, Dordrecht, 465–475, 1999

[Ban99b] N. V. Banichuk. Asymptotic approach to optimal structural design with brittle-fracture constraints (Part 2: Deterministic and stochastic problems). In Mechanics of Composite Materials and Structures, C. A. Mota Soares, et al., eds., Kluwer Academic Publishers, Dordrecht, 477–487, 1999

[BBB$^+$00] N. V. Banichuk, F. J. Barthold, A. I. Borovkov, V. A. Palmov, V. V. Saurin, and E. Stein. Shape optimization of laminated structures under strength constraints, caused by interlayered fracture. In Problems of Strength and Plasticity, Nizhni Novgorod University Press, Cambridge, 62: 19–30, 2000

[BBS05] N. V. Banichuk, F.-J. Barthold, and M. Serra. Optimization of axisymmetric membrane shells against brittle fracture. Meccanica, 40(2): 135–145, 2005

[BH75] A. E. Bryson and Y. C. Ho. Applied Optimal Control. John Wiley & Sons, Inc., New York, 1975

[BIM07] N. V. Banichuk, S. Yu. Ivanova, and E. V. Makeev. Some problems of optimizing shell shape and thickness distribution on the basis of a genetic algorithm. Mechanics of Solids, Allerton Press, 42(6): 956–964, 2007

[BIMS05a] N. V. Banichuk, S. Yu. Ivanova, E. V. Makeev, and A. V. Sinitsin. Optimal shape design of axisymmetric shells for crack initiation and propagation under cyclic loading. Mechanics Based Design of Structures and Machines, 33(2): 253–269, 2005

[BIMS05b] N. V. Banichuk, S. Yu. Ivanova, E. V. Makeev, and A. V. Sinitsin. Several problems of optimal design of shells with damage accumulation taken into account. In Problems of Strength and Plasticity (Izd-vo NNGU), N. Novgorod, 67: 46–59, 2005, in Russian

[BK76] N. V. Banichuk and B. L. Karihaloo. Minimum-weight design of multipurpose cylindrical bars. International Journal of Solids and Structures, 12(4): 267–273, 1976

[Bli46] G. A. Bliss. Lectures on the Calculus of Variations, University of Chicago Press (reprinted by Dover Co), Chicago, Ill. 1946

[BMN00] N. V. Banichuk, M. Mäkelä, and P. Neittaanmäki. Shape optimization for structures from quasi-brittle materials subjected to cyclic loads. In Identification, Control and Optimization of Engineering Structures, G. De Roeck and B. H. V. Topping, eds., CIVIL-COMP Press, Edinburg, 145–151, 2000

[BN07] N. V. Banichuk and P. Neittaanmäki. On structural optimization with incomplete information. Mechanics Based Design of Structures and Machines, 35(1): 75–95, 2007

[BN08a] N. V. Banichuk and P. Neittaanmäki. Incompleteness of information and reliable optimal design. In Evolutionary and Deterministic Methods for Design, Optimization and Control, P. Neittaanmäki, J. Périaux, and T. Tuovinen, eds., CIMNE, Barcelona, pp. 29–38, 2008

[BN08b] N. V. Banichuk and P. Neittaanmäki. Optimal design with incomplete information using worst case scenario. In Advances in Mechanics: Dynamics and Control. Proceedings of the 14th International Workshop on Dynamics and Control, F. L. Chernousko, G. V. Kostin and V. V. Saurin, eds., Nauka, Moscow, 46–52, 2008

[Bol61] V. V. Bolotin. Statistical theory of aseismic design of structures. In Proceeding of Second World Conference on Earthquake Engineering, Tokio, 1961

[Bol69] V. V. Bolotin. Statistical Methods in Structural Mechanics. Holden-Day, Inc., San Francisco, 1969

[BP93] F. J. Baron and O. Pironneau. Multidisciplinary optimal design of a wing profile. In Structural Optimization 93, vol. 2. The World Congress on Optimal Design of Structural Systems (Rio de Janeiro), J. Herskovits, ed., COPPE/Federal University of Rio de Janeiro, Rio de Janeiro, 61–68, 1993

[BPB+00] A. Borovkov, V. Palmov, N. Banichuk, E. Stein, V. Saurin, F. Barthold, and
          Yu. Misnik. Macrofailure criterion and optimization of composite structures with
          edge delamination. International Journal for Computational Civil and Structural
          Engineering, 1(1): 91–104, 2000

[Bra84]   A. M. Brandt, ed., Criteria and Methods of Structural Optimization, Martinus
          Nijhoff Publishers, The Hague, 1984

[BRS03a] N. V. Banichuk, F. Ragnedda, and M. Serra. Some probabilistic problems of beam
          and frame optimization under longevity constraint. Mechanics Based Design of
          Structures and Machines, 31(1): 57–77, 2003

[BRS03b] N. V. Banichuk, F. Ragnedda, and M. Serra. Probabilistic-guaranteed optimal de-
          sign of membrane shells against quasi-brittle fracture. Mechanics Based Design of
          Structures and Machines, 31(4): 459–474, 2003

[BRS06]   N. V. Banichuk, F. Ragnedda, and M. Serra. Axisymmetric shell optimization un-
          der fracture mechanics and geometric constraint. Structural and Multidisciplinary
          Optimization, 31(3): 223–228, 2006. DOI 10.1007/s00158-005-0585-2

[BRS08]   N. V. Banichuk, F. Ragnedda, and M. Serra. Optimization of mass effectiveness
          of axisymmetric pressure vessels. Structural and Multidisciplinary Optimization,
          35(5): 453–459, 2008. DOI 10.1007/s00158-007-0149-8

[BRS99]   N. V. Banichuk, F. Ragnedda, and M. Serra. Probabilistic approaches for optimal
          beam design based on fracture mechanics. Meccanica, 34(1): 29–38, 1999

[BS93]   T. Bäck and H. P. Schwefel. An overview of evolutionary algorithms for parameter
          optimization. Evolutionary Computation, 1(1):1–23, 1993

[BV07]   V. Baranenko and A. Vojnakov. Optimal structural design at random and fuzzy
          information about loading. In Theoretical Foundation of Civil Engineering – XV,
          W. Szczesniak, ed., OW PW, Warsaw, 25–32, 2007

[CB73]   F. L. Chernousko and N. V. Banichuk. Variational Problems of Mechanics and
          Control. Numerical Methods. Nauka, Moscow, 1973, in Russian

[Cea81]   J. Cea. Problems of shape optimal design. In Optimization of Distributed Parameter
          Structures, E. J. Haug and J. Cea, eds., Sijhoff and Noordhoff, Alphen aan den Rijn,
          1005–1048, 1981

[CF88]   G. Cheng and B. Fu. Shape optimization of continuum with crack. In Structural
          Optimization, G. I. N. Rozvany and B. L. Karihaloo, eds., Kluwer Academic
          Publishers, Dordrecht, 57–62, 1988

[Che79]   G. P. Cherepanov. Mechanics of Brittle Fracture. McGraw-Hill, New York, 1979

[CK78]   F. L. Chernousko and V. B. Kolmanovskii. Optimal Control Under Random
          Perturbations. Nauka, Moscow, 1978, in Russian

[CM78]   F. L. Chernousko and A. A. Melikjan. Game-Theoretical Control and Search
          Problems. Nauka, Moscow, 1978, in Russian

[DFH77]   J. W. Davidson, L. P. Felton, and G. C. Hart. Optimum design of structures with
          random parameters, Computers & Structures, 7(3): 481–486, 1977

[DL88]   R. Dautray and J.-L. Lions. Mathematical Analysis and Numerical Methods for
          Science and Technology, Vol. 2, Functional and Variational Methods, Springer-
          Verlag, Berlin, 1988

[DZ01]   M. C. Delfour and J.-P. Zolesio. Shapes and Geometries: Analysis, Differen-
          tial Calculus, and Optimization, Advances in Design and Control 4, SIAM,
          Philadelphia, PA, 2001

[EKO90]   H. Eschenauer, J. Koski, and A. Osyczka, eds., Multicriteria Design Optimization.
          Procedures and Applications. Springer-Verlag, Berlin, 1990

[EKS86]   H. Eschenauer, G. Kneppe, and K. H. Stenvers. Deterministic and stochastic
          multiobjective optimization of beam and shell structures, Journal of Mechanisms,
          Transmission and Automation in Design, 108: 31–37, 1986

[EO83]   H. Eschenauer and N. Olhoff, eds., Optimization Methods in Structural Design,
          Bibliographisches Institut, Mannheim, 1983

[ETB84] E. D. Eason, J. M. Thomas, and P. M. Besuner. Optimization of safety factors and testing plans for structural designs. In New Directions in Optimum Structural Design, E. Atrek, et al. John Wiley & Sons Ltd., Chichester, 505–522, 1984

[ETC96] R. El Abdi, M. Touratier, and P. Convert. Optimal design for minimum weight in a cracked pressure vessel of a turboshaft. Communications in Numerical Methods in Engineering, 12(5): 271–280, 1996

[Flu73] W. Flugge. Stresses in Shells. 2nd ed., Springer-Verlag, Berlin, 1973

[GE94] R. V. Goldstein and V. M. Entov. Qualitative Methods in Continuum Mechanics, Longman, Harlow, copublished with John Wiley & Sons, New York, 1994

[GF63] I. M. Gelfand and S. V. Fomin. Calculus of Variations. Prentice Hall Inc., Englewood Cliffs, NJ, 1963

[Gol89] D. E. Goldberg. Genetic Algorithms in Search, Optimization and Machines Learning, Addison-Wesley, Reading, MA, 1989

[HA79] E. J. Haug and J. S. Arora. Applied Optimal Design, John Wiley & Sons, New York, 1979

[Hau81] E. J. Haug. A review of distributed parameter structural optimization literature. In Optimization of Distributed Parameter Structures, Vol. 1, E. J. Haug and J. Cea, eds., Sijthoff and Noordhoff, Alphen aan den Rijn, 3–74, 1981

[HC81] E. J. Haug and J. Cea, eds., Optimization of Distributed Parameter Structures, Vol. 1 and Vol. 2, Sijthoff and Noordhoff, Alphen aan den Rijn, 1981

[HCK86] E. J. Haug, K. K. Choi, and V. Komkov. Design Sensitivity Analysis of Structural Systems, Academic Press, Orlando, FL, 1986

[Hel84] K. Hellan. Introduction to Fracture Mechanics, Mc Graw-Hill Inc., New York, 1984

[HG92] R. T. Haftka and Z. Gürdal. Elements of Structural Optimization, 3rd ed., Kluwer Academic Publishers, Dordrecht, 1992

[HGK90] R. T. Haftka, Z. Gürdal, and M. P. Kamat. Elements of Structural Optimization, 2nd ed., Kluwer Academic Publishers, Dordrecht, 1990

[HJK⁺00] J. Haslinger, D. Jedelsky, T. Kozubek, and J. Tvrdik. Genetic and random search methods in optimal shape design problems, Journal of Global Optimization. 16(2): 109–131, 2000

[HM03] J. Haslinger and R. A. E. Mäkinen. Introduction to Shape Optimization. Theory, Approximation, and Computation, SIAM, Philadelphia, PA, 2003

[HN88] J. Haslinger and P. Neittaanmäki. Finite Element Approximation for Optimal Shape Design. Theory and Applications. John Wiley & Sons, Chichester, 1988

[HN96] J. Haslinger and P. Neittaanmäki. Finite Element Approximation for Optimal Shape, Material and Topology Design. John Wiley & Sons, Chichester, 2nd edition, 1996

[HNT86] J. Haslinger and P. Neittaanmäki, and T. Tiihonen. Shape optimization in contact problems based on penalization of the state inequality, Aplikace Matematiky, 31(1): 54–77, 1986

[Hol75] J. H. Holland. Adaptation in Neural and Artificial Systems, University of Michigan Press, Ann Arbor, MI, 1975

[Hut79] J. W. Hutchinson. A Course of Nonlinear Fracture Mechanics, Technical University of Denmark, Copenhagen, Lyngby, 1979

[Ing93] L. Ingber. Simulated annealing: Practice versus theory, Mathematical and Computer Modelling, 18(11): 29–57, 1993

[JM83] S. Jendo and W. Marks. Nonlinear stochastic programming in optimum structural design. In Optimization Methods in Structural Design, H. Eschenauer and N. Olhoff, eds., Bibliographisches Institut, Monnheim, 327–333, 1983

[Kar79a] B. L. Karihaloo. Optimal design of multi-purpose structures. Journal of Optimization Theory and Applications, 27(3): 449–461, 1979

[Kar79b] B. L. Karihaloo. Optimal design of multi-purpose tie column of solid construction. International Journal of Solids and Structures, 15(2): 103–109, 1979

[Kom84] V. Komkov, ed., Sensitivity of Functionals with Applications to Engineering Sciences, Lecture Notes in Mathematics 1086, Springer-Verlag, New York, 1988

[Kom88a] V. Komkov. Variational Principles of Continuum Mechanics with Engineering Applications. Vol. 1: Critical Points Theory, Reidel Publishing Co., Dordrecht, 1988

[Kom88b] V. Komkov. Variational Principles of Continuum Mechanics with Engineering Applications. Vol. 2: Introduction to Optimal Design Theory, Reidel Publishing Co., Dordrecht, 1988

[KP79] B. L. Karihaloo and R. D. Parbery. Optimal design of multi-purpose beam-columns. Journal of Optimization Theory and Applications, 27(3): 439–448, 1979

[KP80] B. L. Karihaloo and R. D. Parbery. Optimal design of beam-columns subjected to concentrated moments. Engineering Optimization, 5(1): 59–65, 1980

[KP85] M. F. Kanninen and C. H. Popelar. Advanced Fracture Mechanics, Oxford University Press, New York, 1985

[Kur77] A. B. Kurzhanskii. Control and Observation under Conditions of Indeterminacy. Nauka, Moscow, 1977, in Russian

[LL50] M. A. Lavrentyev and L. A. Lusternik. Course of the Calculus of Variations. Gostekhteorizdat, Moscow, 1950, in Russian

[LM96] A. K. Lyubimov and F. V. Makarenko. Probability approach to the optimization problem of a reinforced plate with a crack. In Applied Problems of Strength and Plasticity, KMK Scientific Press, Moscow, 120–131, 1996

[Lov44] A. E. H. Love. A Treatise on the Mathematical Theory of Elasticity. Dover Publications, New York, 4th edition, 1944

[Mar70] J. B. Martin. Optimal Design of elastic structures for multipurpose loading, Journal of Optimization Theory and Applications, 6(1): 22–40, 1970

[Mar95] K. Marti. Derivatives of probability functions arising in probabilistic structural analysis and design. In First World Congress of Structural and Multidisciplinary Optimization, N. Olhoff and G. I. N. Rozvany, eds., Pergamon, Redwood Books, Trowbridge, Great Britain, 877–882, 1995

[Mie99] K. Miettinen. Nonlinear Multiobjective Optimization, Kluwer Academic Publishers, Boston, MA, 1999

[MMM99] J. Mäkinen, K. Miettinen, and M. Mäkelä. Some penalty methods with genetic algorithms. In Proceedings of EUGOGEN99 – Short Course on Evolutionary Algorithms in Engineering and Computer Science, K. Miettinen, M. Mäkelä, and J. Toivanen, eds., Reports of the Department of Mathematical Information Technology, Ser. A, Collections, A2/1999, Univ. of Jyväskylä, Jyväskylä, 105–112, 1999

[MMNP99] K. Miettinen, M. M. Mäkelä, P. Neittaanmäki, and J. Periaux, eds., Evolutionary Algorithms in Engineering and Computer Science. John Wiley & Sons, Chichester, 1999

[Mor84] N. F. Morozov. Mathematical Problems of Fracture Theory. Nauka, Moscow, 1984, in Russian

[Mos77] F. Moses. Structural system reliability and optimization, Computers & Structures, 7(2): 283–290, 1977

[MU81] V. P. Malkov and A. G. Ugodchikov. Optimization of Elastic Systems, Nauka, Moscow, 1981

[Mus53] N. I. Muskhelishvili. Some Basic Problems of the Mathematical Theory of Elasticity. Noordhoff, Groningen, 1953

[Nei91] P. Neittaanmäki. Computer aided optimal structural design. Surveys on Mathematics for Industry, 1: 173–215, 1991

[Now70] W. Nowacki. Teoria Sprezystosci, Panstwowe Wydawnictwo Naukowe, Warszawa, 1970

[NR04] P. Neittaanmäki and S. Repin. Reliable Methods for Computer Simulation. Error Control and a Posteriori Estimates, Studies in Mathematics and its Applications, vol. 33, Elsevier, Amsterdam, 2004

[NST06] P. Neittaanmäki, J. Sprekels, and D. Tiba. Optimization of Elliptic Systems. Theory and Applications, Springer, New York, 2006

[OR95]   N. Olhoff and G. I. N. Rozvany, eds., First World Congress of Structural and Multidisciplinary Optimization (WCS MO - 1), Pergamon, Redwood Books, Trowbridge, Great Britain, 1995

[Par92] V. Z. Parton. Fracture Mechanics – From Theory to Practice, Gordon and Breach Science Publishers, Philadelphia, 1992

[PH03]   M. Papila and R. T. Haftka. Implementation of a crack propagation constraint within a structural optimization software. Structural and Multidisciplinary Optimization, 25(5–6): 327–338, 2003

[Pic88] A. Picuda. Optimality conditions for multiply loaded structures – integrated control and finite element method. In Structural Optimization, G. I. N. Rozvany and B. L. Karihaloo, eds., Kluwer Academic Publishers, Dordrecht, 233–239, 1988

[PJPO01] J. Periaux, P. Joly, O. Pironneau, and E. Onate, eds., Innovative Tools for Scientific Computation in Aeronautical Engineering. CIMNE, a Series of Handbooks on Theory and Engineering Applications of Computational Methods, Barcelona, 2001

[PK77]   R. D. Parbery and B. L. Karihaloo. Minimum-weight design of hollow cylinders for given lower bounds on torisonal and flexural rigidities. International Journal of Solids and Structures, 13(12): 1271–1280, 1977

[PK80]   R. D. Parbery and B. L. Karihaloo. Minimum-weight design of thin-walled cylinders subject to flexural and torsional stiffness constraints, Journal of Applied Mechanics, 47(1): 106–110, 1980

[PKK97] E. Polak, C. Kijner-Neto, and A. Der Kiureghian. Structural optimization with reliability constrains. In Reliability and Optimization of Structural Systems, D. M. Frangopol, R. B. Corotis, and R. Rackwitz, eds., Pergamon, Redwood Books, Trowbridge, 17–32, 1997

[Pol48] G. Polya. Torsional rigidity, principal frequency, electrostatic capacity and symmetrization. Quarterly of Applied Mathematics, 6(3): 267–277, 1948

[Pra72] W. Prager. Introduction to Structural Optimization, Springer-Verlag, Wien, 1972

[PS51]   G. Polya and G. Szegö. Isoperimetric Inequalities in Mathematical Physics. Princeton University Press, Princeton, NJ, 1951

[PS68]   W. Prager and R. T. Shield. Optimal design of multi-purpose structures. International Journal of Solids and Structures, 4(4): 469–475, 1968

[QPPW98] D. Quagliarella, J. Periaux, C. Poloni, and G. Winter, eds., Genetic Algorithms and Evolution Strategies in Engineering and Computer Science. Recent Advances and Industrial Applications. John Wiley & Sons, Chichester, 1998

[RK88]   G. I. N. Rozvany and B. L. Karihaloo, eds., Structural Optimization, Kluwer Academic Publishers, Dordrecht, 1988

[Roz76] G. I. N. Rozvany. Optimal Design of Flexural Systems, Pergamon, Oxford, 1976

[Ser00] M. Serra. Optimum beam design based on fatigue crack propagation. Structural and Multidisciplinary Optimization, 19(2): 159–163, 2000

[SK89]   W. Stadler and V. Krishnan. Natural structural shapes for shells of revolution in the membrane theory of shells. Structural and Multidisciplinary Optimization, 1(1): 19–27, 1989

[SL68]   G. H. Sih and H. Liebowitz. Mathematical theory of brittle fracture. In Fracture, H. Liebowitz, ed., vol.2, Mathematical Fundamentals, Academic Press, New York, 1968

[Smi91] R. N. L. Smith. BASIC Fracture Mechanics: Including an Introduction to Fatigue. Butterworth-Heinemann Ltd., Oxford, 1991

[Str47] S. Streletskii. Problem of establishing safety factors for structures, Izvestiya Akademii Nauk SSSR, No. 1, 1947

[Tim56] S. Timoshenko. Strength of Materials. Part II, Advanced Theory and Problems, Van Nostrag Co., New York, 1956

[Tim87] S. P. Timoshenko. Theory of Elasticity. McGraw-Hill, New York, 3rd edition, 1987

[TW59]   S. Timoshenko and S. Woinowsky-Krieger. Theory of Plates and Shells, McGraw-Hill, New York, 1959

[TWK94] N. B. Thomsen, J. Wang, and B. L. Karihaloo. Optimization – a tool in advanced material technology. Structural and Multidisciplinary Optimization, 8(1): 9–15, 1994

[VHS02] R. Vitali, R. T. Haftka, and B. V. Sankar. Multi-fidelity design of stiffened composite panel with a crack. Structural and Multidisciplinary Optimization, 23(5): 347–356, 2002

[Was82] K. Washizu. Variational Methods in Elasticity and Plasticity, Pergamon, Oxford, 1982

[WK95] J. Wang and B. L. Karihaloo. Fracture mechanics and optimization – a useful tool for fibre-reinforced composite design. Composite Structures, 32(1–4): 453–466, 1995

[YCC97] X. Yu, K. K. Choi, and K. H. Chang. A mixed design approach for probabilistic structural durability, Structural and Optimization, 14(2–3): 81–90, 1997

Index